Anonymous

The Opium Habit

With Suggestions as to the Remedy

Anonymous

The Opium Habit
With Suggestions as to the Remedy

ISBN/EAN: 9783337805463

Printed in Europe, USA, Canada, Australia, Japan

Cover: Foto ©berggeist007 / pixelio.de

More available books at **www.hansebooks.com**

SUGGESTIONS AS TO THE REMEDY.

"It is almost like Dives seeking for a messenger to his brethren; but tell them —tell all *young men* what it is, 'that they come not into this torment.' "—*Request of a dying opium patient.*

NEW YORK:

HARPER & BROTHERS, PUBLISHERS,

FRANKLIN SQUARE.

1868.

" After my death, I earnestly entreat that a full and unqualified narrative of my wretchedness, and of its guilty cause, may be made public, that at least some little good may be effected by the direful example."—COLERIDGE.

CONTENTS.

	PAGE
INTRODUCTION	5
A SUCCESSFUL ATTEMPT TO ABANDON OPIUM	11
DE QUINCEY'S "CONFESSIONS OF AN ENGLISH OPIUM-EATER"	77
OPIUM REMINISCENCES OF COLERIDGE	133
WILLIAM BLAIR	179
OPIUM AND ALCOHOL COMPARED	198
INSANITY AND SUICIDE FROM AN ATTEMPT TO ABANDON MOR-PHINE	224
A MORPHINE HABIT OVERCOME	232
ROBERT HALL—JOHN RANDOLPH—WILLIAM WILBERFORCE	240
WHAT SHALL THEY DO TO BE SAVED?	250
OUTLINES OF THE OPIUM-CURE	285

INTRODUCTION.

THIS volume has been compiled chiefly for the benefit of opium-eaters. Its subject is one indeed which might be made alike attractive to medical men who have a fancy for books that are professional only in an accidental way ; to general readers who would like to see gathered into a single volume the scattered records of the consequences attendant upon the indulgence of a pernicious habit ; and to moralists and philanthropists to whom its sad stories of infirmity and suffering might be suggestive of new themes and new objects upon which to bestow their reflections or their sympathies. But for none of these classes of readers has the book been prepared. In strictness of language little medical information is communicated by it. Incidentally, indeed, facts are stated which a thoughtful physician may easily turn to professional account. The literary man will naturally feel how much more attractive the book might have been made had these separate and sometimes disjoined threads of mournful personal histories been woven into a more coherent whole ; but the book has not been made for literary men. The philanthropist, whether a theoretical or a practical one, will find in its pages little preaching after his particular vein, either upon the vice or the danger of opium-eating. Possibly, as he peruses these various records, he may do much preaching for himself, but he will not find a great deal furnished to his hand, always excepting the rather inopportune reflections of Mr. Joseph Cottle over the case of his unhappy friend Coleridge. The book has been compiled for opium-eaters, and to their notice it is urgently commended.

Sufferers from protracted and apparently hopeless disorders profit little by scientific information as to the nature of their complaints, yet they listen with profound interest to the experience of fellow-sufferers, even when this experience is unprofessionally and unconnectedly told. Medical empirics understand this and profit by it. In place of the general statements of the educated practitioner of medicine, the empiric encourages the drooping hopes of his patient by narrating in detail the minute particulars of analagous cases in which his skill has brought relief.

Before the victim of opium-eating is prepared for the services of an intelligent physician he requires some stimulus to rouse him to the possibility of recovery. It is not the *dicta* of the medical man, but the experience of the relieved patient, that the opium-eater, desiring — nobody but he knows how ardently—to enter again into the world of hope, needs, to quicken his paralyzed will in the direction of one tremendous effort for escape from the thick night that blackens around him. The confirmed opium-eater is habitually hopeless. His attempts at reformation have been repeated again and again ; his failures have been as frequent as his attempts. He sees nothing before him but irremediable ruin. Under such circumstances of helpless depression, the following narratives from fellow-sufferers and fellow-victims will appeal to whatever remains of his hopeful nature, with the assurance that others who have suffered even as he has suffered, and who have struggled as he has struggled, and have failed again and again as he has failed, have at length escaped the destruction which in his own case he has regarded as inevitable.

The number of confirmed opium-eaters in the United

States is large, not less, judging from the testimony of druggists in all parts of the country as well as from other sources, than eighty to a hundred thousand. The reader may ask who make up this unfortunate class, and under what circumstances did they become enthralled by such a habit? Neither the business nor the laboring classes of the country contribute very largely to the number. Professional and literary men, persons suffering from protracted nervous disorders, women obliged by their necessities to work beyond their strength, prostitutes, and, in brief, all classes whose business or whose vices make special demands upon the nervous system, are those who for the most part compose the fraternity of opium-eaters. The events of the last few years have unquestionably added greatly to their number. Maimed and shattered survivors from a hundred battle-fields, diseased and disabled soldiers released from hostile prisons, anguished and hopeless wives and mothers, made so by the slaughter of those who were dearest to them, have found, many of them, temporary relief from their sufferings in opium.

There are two temperaments in respect to this drug. With persons whom opium violently constricts, or in whom it excites nausea, there is little danger that its use will degenerate into a habit. Those, however, over whose nerves it spreads only a delightful calm, whose feelings it tranquillizes, and in whom it produces an habitual state of reverie, are those who should be upon their guard lest the drug to which in suffering they owe so much should become in time the direst of curses. Persons of the first description need little caution, for they are rarely injured by opium. Those of the latter class, who have already become enslaved by the habit, will find many things in these pages that are in harmony with their own experience; other things they will doubtless find of which they have had no experience. Many of the particular effects of opium differ according to the different constitutions of those who use it. In De Quincey it exhibited its power in gorgeous dreams in consequence of some special tendency in that direction in De Quincey's temperament, and not because dreaming is by any means an invariable attendant upon opium-eating. Different races also seem to be differently affected by its use. It seldom, perhaps never, intoxicates the European ; it seems habitually to intoxicate the Oriental. It does not generally distort the person of the English or American opium-eater ; in the East it is represented as frequently producing this effect.

It is doubtful whether a sufficient number of cases of excess in opium-eating or of recovery from the habit have yet been recorded, or whether such as have been recorded have been so collated as to warrant a positive statement as to all the phenomena attendant upon its use or its abandonment. A competent medical man, uniting a thorough knowledge of his profession with educated habits of generalizing specific facts under such laws—affecting the nervous, digestive, or secretory system—as are recognized by medical science, might render good service to humanity by teaching us properly to discriminate in such cases between what is uniform and what is accidental. In the absence, however, of such instruction, these imperfect, and in some cases fragmentary, records of the experience of opium-eaters are given, chiefly in the language of the sufferers themselves, that the opium-eating reader may compare case with case, and deduce from such comparison the lesson of the entire practicability of his own release from what has been the burden and the curse of his existence. The entire object of the compilation will have been attained, if the narratives given in these pages shall be found to serve the double purpose of indicating to the beginner in opium-eating the hazardous path he is treading, and of awakening in the confirmed victim of the habit the hope that he may be released from the frightful thraldom which has so long held him, infirm in body, imbecile in will, despairing in the present, and full of direful foreboding for the future.

In giving the subjoined narratives of the experience of opium-eaters, the compiler has been sorely tempted to weave them into a more coherent and connected story ; but he has been restrained by the conviction that the thousands of opium-eaters, whose relief has been his main object in preparing the volume, will be more benefited by allowing each sufferer to tell his own story than by any attempt on his part to generalize the multifarious and often discordant phenomena attendant upon the disuse of opium. As yet the medical profession are by no means agreed as to the character or proper treatment of the opium disease. While medical science remains in this state, it would be impertinent in any but a professional person to attempt much more than a statement of his own case, with such general advice as would naturally occur to any intelligent sufferer. Very recently indeed, some suggestions for the more successful treatment of the habit have been discussed both by eminent medical men and by distinguished philanthropists. Could an Institution for this purpose be established, the chief difficulty in the way of the redemption of unhappy thousands would be obviated. The general outline of such a plan will be found at the close of the volume. It seems eminently deserving the profound consideration of all who devote themselves to the promotion of public morals or the alleviation of individual suffering.

THE OPIUM HABIT.

A SUCCESSFUL ATTEMPT TO ABANDON OPIUM.

IN the personal history of many, perhaps of most, men, some particular event or series of events, some special concurrence of circumstances, or some peculiarity of habit or thought, has been so unmistakably interwoven and identified with their general experience of life as to leave no doubt in the mind of any one of the decisive influence which such causes have exerted. Unexaggerated narrations of marked cases of this kind, while adding something to our knowledge of the marvellous diversities of temptation and trial, of success and disappointment which make up the story of human life, are not without a direct value, as furnishing suggestions or cautions to those who may be placed in like circumstances or assailed by like temptations.

The only apology which seems to be needed for calling the attention of the reader to the details which follow of a violent but successful struggle with the most inveterate of all habits, is to be found in the hope which the writer indulges, that while contributing something to the current amount of knowledge as to the horrors attending the habitual use of opium, the story may not fail to encourage some who now regard themselves as hopeless victims of its power to a strenuous and even desperate effort for recovery. Possibly the narrative may also not be without use to those who are now merely in danger of becoming enslaved by opium, but who may be wise enough to profit in time by the experience of another.

A man who has eaten much more than half a hundred-weight of opium, equivalent to more than a hogshead of laudanum, who has taken enough of this poison to destroy many thousand human lives, and whose uninterrupted use of it continued for nearly fifteen years, ought to be able to say something as to the good and the evil there is in the habit. It forms, however, no part of my purpose to do this, nor to enter into any detailed statement of the circumstances under which the habit was formed. I neither wish to diminish my own sense of the evil of such want of firmness as characterizes all who allow themselves to be betrayed into the use of a drug which possesses such power of tyrannizing over the most resolute will, nor to withdraw the attention of the reader from the direct lesson this record is designed to convey, by saying any thing that shall seem to challenge his sympathy or forestall his censures. It may, however, be of service to other opium-eaters for me to state briefly, that while endowed in most respects with uncommon vigor of constitution, and with a temperament singularly remote from any tendency to despondency or hypochondria, an unusual nervous sensibility, together with a constitutional tendency to a disordered condition of the digestive organs, strongly predisposed me to accept the fascination of the opium habit. The difficulty, early in life, of retaining food of any kind upon the stomach was soon followed by vagrant shooting pains over the body, which at a later day assumed a permanent chronic form.

After other remedies had failed, the eminent physician under whose advice I was acting recommended opium. I have

no doubt he acted both wisely and professionally in the prescription he ordered, but where is the patient who has learned the secret of substituting luxurious enjoyment in place of
acute pain by day and restless hours by night, that can be
trusted to take a correct measure of his own necessities?
The result was as might have been anticipated : opium after
a few months' use became indispensable. With the full consciousness that such was the case, came the resolution to
break off the habit. This was accomplished after an effort
no more earnest than is within the power of almost any one
to make. A recurrence of suffering more than usually severe led to a recourse to the same remedy, but in largely increased quantities. After a year or two's use the habit was
a second time broken by another effort much more protracted and obstinate than the first. Nights made weary and
days uncomfortable by pain once more suggested the same
unhappy refuge, and after a struggle against the supposed
necessity, which I now regard as half-hearted and cowardly,
the habit was resumed, and owing to the peculiarly unfavorable state of the weather at the time, the quantity of opium
necessary to alleviate pain and secure sleep was greater than
ever. The habit of relying upon large doses is easily established ; and, once formed, the daily quantity is not easily reduced. All persons who have long been accustomed to
opium are aware that there is a *maximum* beyond which
no increase in quantity does much in the further alleviation
of pain or in promoting increased pleasurable excitement.
This maximum in my own case was eighty grains, or two
thousand drops of laudanum, which was soon attained, and
was continued, with occasional exceptions, sometimes dropping below and sometimes largely rising above this amount,
down to the period when the habit was finally abandoned.
I will not speak of the repeated efforts that were made dur-

ing these long years to relinquish the drug. They all failed, either through the want of sufficient firmness of purpose, or from the absence of-sufficient bodily health to undergo the suffering incident to the effort, or from unfavorable circumstances of occupation or situation which gave me no adequate leisure to insure their success. At length a concurrence of favorable circumstances induced me to resolve upon a final effort to emancipate myself from the habit.   .

For two or three years previous to this time my general health had been gradually improving. Neuralgic disturbance was of less frequent occurrence and was less intense, the stomach retained its food, and, what was of more consequence, the difficulty of securing a reasonable amount of sleep had for the most part passed away. Instead of a succession of wakeful nights any serious interruption of habitual rest occurred at infrequent intervals, and was usually limited to a single night.

In addition to these hopeful indications in encouragement of a vigorous effort to abandon the habit, there were on the other hand certain warnings which could not safely be neglected. The stomach began to complain,—as well it might after so many years unnatural service,—that the daily task of disposing of a large mass of noxious matter constantly cumulating its deadly assaults upon the natural processes of life was getting to be beyond its powers. The pulse had become increasingly languid, while the aversion to labor of any kind seemed to be settling down into a chronic and hopeless infirmity. Some circumstances connected with my own situation pointed also to the appropriateness of the present time for an effort which I knew by the experience of others would make a heavy demand upon all one's fortitude, even when these circumstances were most propitious. At this period my time was wholly at my own disposal. My family was a small one, and I was sure of every accessory support I might need from them to tide me over what I hoped would prove only a temporary, though it might be a severe, struggle. The house I occupied was fortunately so situated that no outcry of pain, nor any extorted eccentricity of conduct, consequent upon the effort I proposed to make, could be observed by neighbors or by-passers.

A few days before the task was commenced, and while on a visit to the capital of a neighboring State in company with a party of gentlemen from Baltimore, I had ventured upon reducing by one-quarter the customary daily allowance of eighty grains. Under the excitement of such an occasion I continued the experiment for a second day with no other perceptible effect than a restless indisposition to remain long in the same position. This, however, was a mere experiment, a prelude to the determined struggle I was resolved upon making, and to which I had been incited chiefly through the encouragement suggested by the success of De Quincey. There is a page in the "Confessions" of this author which I have no doubt has been perused with intense interest by hundreds of opium-eaters. It is the page which gives in a tabular form the gradual progress he made in diminishing the daily quantity of laudanum to which he had long been accustomed. I had read and re-read with great care all that he had seen fit to record respecting his own triumph over the habit. I knew that he had made use of opium irregularly and at considerable intervals from the year 1804 to 1812, and that during this time opium had not become a daily necessity; that in the year 1813 he had become a confirmed opium-eater, "of whom to ask whether on any particular day he had or had not taken opium, would be to ask whether his lungs had performed respiration, or the heart

fulfilled its functions;" that in the year 1821 he had published his " Confessions," in which, while leading the unob-
servant reader to think that he had mastered the habit, he
had in truth only so far succeeded as to reduce his daily allowance from a quantity varying from fifty or sixty to one
hundred and fifty grains, down to one varying from seven to
twelve grains ; that in the year 1822 an appendix was added
to the " Confessions " which contained a tabular statement
of his further progress toward an absolute abandonment of
the drug, and indicating his gradual descent, day by day, for
thirty-five days, when the reader is naturally led to suppose
that the experiment was triumphantly closed by his entire
disuse of opium.

I had failed, however, to observe that a few pages preceding this detailed statement the writer had given a faint inti-
mation that the experiment had been a more protracted one
than was indicated by the table.   I had also failed to notice
the fact that no real progress had been made during the
first four weeks of the attempt : the average quantity of
laudanum daily consumed for the first week being one hundred and three drops ; of the second, eighty-four drops ; of
the third, one hundred and forty-two drops ; and of the
fourth, one hundred and thirty-eight drops ; and that in the
fifth week the self-denial of more than three days had been
rewarded with the indulgence of three hundred drops on the
fourth.   A careful comparison of this kind, showing that in
an entire month the average of the first week had been but
one hundred and three drops, while the average of the last
had been one hundred and thirty-eight drops, and that in
the fifth week a frantic effort to abstain wholly for three
days had obliged him to use on the fourth more than double
the quantity to which of late he had been accustomed, would
have prevented the incautious conclusion, suggested by his

table, that De Quincey made use of laudanum but on two occasions after the expiration of the fourth week.

Whatever may have been the length of time taken by De Quincey "in unwinding to its last link the chain which bound" him, it is certain we have no means of knowing it from any thing he has recorded. Be it shorter or longer, his failure to state definitely the entire time employed in his experiment occasioned me much and needless suffering. I thought that if another could descend, without the experience of greater misery than De Quincey records, from one hundred and thirty drops of laudanum, equivalent to about five grains of opium, to nothing, in thirty-four or five days, and in this brief period abandon a habit of more than nine years' growth, a more resolved will might achieve the same result in the same number of days, though the starting-point in respect to aggregate quantity and to length of use was much greater. The object, therefore, to be accomplished in my own case was to part company forever with opium in thirty-five days, cost what suffering it might. On the 26th of November, in a half-desperate, half-despondent temper of mind, I commenced the long-descending *gradus* which I had rapidly ascended so many years before. During this entire period the quantity consumed had been pretty uniformly eighty grains of best Turkey opium daily. Occasional attempts to diminish the quantity, but of no long continuance, and occasional overindulgence during protracted bad weather, furnished the only exceptions to the general uniformity of the habit.

The experiment was commenced by a reduction the first day from eighty grains to sixty, with no very marked change of sensations ; the second day the allowance was fifty grains, with an observable tendency toward restlessness, and a general uneasiness ; the third day a further reduction of ten grains had diminished the usual allowance by one-half, but with a perceptible increase in the sense of physical discomfort. The mental emotions, however, were entirely jubilant. The prevailing feeling was one of hopeful exultation. The necessity for eighty grains daily had been reduced to a necessity for only forty, and, therefore, one-half of the dreaded task seemed accomplished. It was a great triumph, and the remaining forty grains were a mere *bagatelle*, to be disposed of with the same serene self-control that the first had been. A weight of brooding melancholy was lifted from the spirits : the world wore a happier look. The only drawback to this beatific state of mind was a marked indisposition to remain quiet, and a restless aversion to giving attention to the most necessary duties.

Two days more and I had come down to twenty-five grains. Matters now began to look a good deal more serious. Only fifteen of the last forty grains had been dispensed with ; but this gain had cost a furious conflict. A strange compression and constriction of the stomach, sharp pains like the stab of a knife beneath the shoulder-blades, perpetual restlessness, an apparent prolongation of time, so much so that it seemed the day would never come to a close, an incapacity of fixing the attention upon any subject whatever, wandering pains over the whole body, the jaw, whenever moved, making a loud noise, constant irritability of mind and increased sensibility to cold, with alternations of hot flushes, were some of the phenomena which manifested themselves at this stage of the process. The mental elations of the first three days had become changed by the fifth into a state of high nervous excitement ; so that while on the whole there was a prevailing hopefulness of temper, and even some remaining buoyancy of spirits, arising chiefly from the certainty that already the quantity consumed had been reduced by more than two-thirds, the conviction had, nevertheless, greatly deepened, that the task was like to prove a much more serious one than I had anticipated. Whether it was possible at present to carry the descent much further had become a grave question. The next day, however, a reduction of five grains was somehow attained; but it was a hard fight to hold my own within this limit of twenty grains. From this stage commenced the really intolerable part of the experience of an opium-eater retiring from service. During a single week, three-quarters of the daily allowance had been relinquished, and in this fact, at least, there was some ground for exultation. If what had been gained could only be secured beyond any peradventure of relapse, so far a positive success would be achieved.

Had the experiment stopped here for a time until the system had become in some measure accustomed to its new habits, possibly the misery I subsequently underwent might some of it have been spared me. However this may be, I had not the patience of mind necessary for a protracted experiment. What I did must be done at once; if I would win I must fight for it, and must find the incentive to courage in the conscious desperation of the contest.

From the point I had now reached until opium was wholly abandoned, that is, for a month or more, my condition may be described by the single phrase, intolerable and almost unalleviated wretchedness. Not for a waking moment during this time was the body free from acute pain; even in sleep, if that may be called sleep which much of it was little else than a state of diminished consciousness, the sense of suffering underwent little remission. What added to the aggravation of the case, was the profound conviction that no further effort of resolution was possible, and that every counteracting influence of this kind had been already wound up to its highest tension. I might hold my own; to do any thing more I thought impossible. Before the month had come to an end, however, I had a good deal enlarged my conceptions of the possible resources of the will when driven into a tight corner.

The only person outside of my family to whom I had confided the purpose in which I was engaged was a gentleman with whom I had some slight business relations, and who I knew would honor any demands I might make in the way of money. I had assured him that by New Year's Day I should have taken opium for the last time, and that any extravagance of expenditure would not probably last beyond that date. Upon this assurance, but confessedly having little or no faith in it, he asked me to dine with him on the auspicious occasion.

So uncomfortable had my condition and feelings become in the rapid descent from eighty grains to twenty in less than a week, that I determined for the future to diminish the quantity by only a single grain daily, until the habit was finally mastered. In the twenty-nine days which now remained to the first of January, the nine days more than were needed, at the proposed rate of diminution, would, I thought, be sufficient to meet any emergency which might arise from occasional lapses of firmness in adhering to my self-imposed task, and more especially for the difficulties of the final struggle—difficulties I believe to be almost invariably incident to any strife which human nature is called upon to make in overcoming not merely an obstinate habit but the fascination of a long-entranced imagination. Up to this time I had taken the opium as I had always been accustomed to do, in a single dose on awaking in the morning. I now, however, divided the daily allowance into two portions, and after a day or two into four, and then into single grains. The chief advantage which followed this subdivision of the

dose was a certain relief to the mind, which for a few days
had become fully aware of the power which misery possesses
of lengthening out the time intervening between one alleviation and another, and which shrunk from the weary continuance of an entire day's painful and unrelieved abstinence
from the accustomed indulgence. The first three days
from the commencement of this grain by grain descent was
marked by obviously increased impatience with any thing
like contradiction or opposition, by an absolute aversion to
reading, and by a very humiliating sense of the fact that the
vis vitæ had somehow become pretty thoroughly eliminated
from both mind and body. Still, when night came, as with
long-drawn steps it did come, there was the consciousness
that something had been gained, and that this daily gain,
small as it was, was worth all it had cost. The tenth day of
the experiment had reduced my allowance to sixteen grains.
The effect of this rapid diminution of quantity was now made
apparent by additional symptoms. The first tears extorted
by pain since childhood were forced out as by some gland-
ular weakness. Restlessness, both of body and mind, had
become extreme, and was accompanied with a hideous and
almost maniacal irritability, often so plainly without cause as
sometimes to provoke a smile from those who were about
me.

For a few days a partial alleviation from too minute attention to the pains of the experiment were found in vigorous horseback exercise. The friend to whose serviceable-
ness in pecuniary matters I have already alluded, offered me
the use of a saddle-horse. The larger of the two animals
which I found in his stable was much too heroic in appearance for me in my state of exhaustion to venture upon. Besides this, his Roman nose and severe gravity of aspect
somehow reminded me, whenever I entered his stall, of the

late Judge ——, to whose Lectures on the Constitution I had listened in my youth, and in my then condition of moral humiliation I felt the impropriety of putting the saddle on an animal connected with such respectable associations. No such scruples interfered with the use of the other animal, which was kept chiefly, I believe, for servile purposes. He was small and mean-looking — his foretop and mane in a hopeless tangle, with hay-seed on his eyelids, and damp straws scattered promiscuously around his body.

Inconsiderable as this animal was, both in size and action, he was almost too much for me, in the weak state to which I was now reduced. This much, however, I owe him; disreputable-looking as he was, he was still a something upon which my mind could rest as a point of diversion from myself—a something outside of my own miseries. At this time the sense of physical exhaustion had become so great that it required an effort to perform the most common act. The business of dressing was a serious tax upon the energies. To put on a coat, or draw on a boot, was no light labor, and was succeeded by such a feeling of prostration as required considerable time to recover from it. It was usually late in the morning before I could master sufficient energy to venture upon the needed exercise. The distance to my friend's stable was trifling. Sometimes I would find there the negro man to whose care the horses were entrusted, but more frequently he was absent. A feeling of humiliation at being seen by any one at a loss how to mount a horse of so diminutive proportions, would triumph over the sense of bodily weakness whenever he was present to bridle and saddle him. Whenever he was not at hand the task of getting the saddle on the pony's back was a long and arduous one. As for lifting it from its hook and throwing it to its place, I could as easily have thrown the horse itself over the stable. The only way in which it could be effected was by first pushing the saddle from its hook, checking its fall to the floor by the hand, and then resting till the violent action of the heart had somewhat abated; next, with occasional failures, to throw it over the edge of the low manger; then an interval of panting rest. Shortening the halter so far as to bring the pony's head close to the manger, next enabled me easily to push him into a line nearly parallel with it, leaving me barely space enough to pass between. By lengthening the stirrup strap I was enabled to get it across his neck, and by much pulling, finally haul the saddle to its proper place. By a kind of desperation of will I commonly succeeded, though by no means always. Sometimes the mortification and rage at a failure so contemptible assured success on a second trial, with apparently less expenditure of exertion than at first. Occasionally, however, I was forced to call for assistance from sheer exhaustion. The bridling was comparatively an easy matter; with his head so closely tied to the manger little scope was left for dodging. In the irritable condition I was now in, the most trifling opposition made me angry, and anger gave me strength; and in this sudden vigor of mind the issue of our daily struggle was, I believe, with a single exception, on my side.

When I led him into the yard, the insignificance of his appearance, in-contrast with the labor it had cost me to get him there, was enough to make any one laugh, excepting perhaps a person suffering the punishment I was then undergoing. Mounting the animal called for a final struggle of determination with weakness. A stone next the fence was the chief reliance in this emergency. It placed me nearly on a level with the stirrup, while the fence enabled me to steady myself with my hand and counteract the tremulousness of

B

the knees, which made mounting so difficult. On one occasion, however, my dread of being observed induced me to make too great an effort. Hearing some one approach, I attempted to raise myself in the stirrup without the aid of stone or fence, but it was more than I could manage. Hardly had I succeeded in raising myself from the ground when my extreme feebleness was manifest, and I fell prostrate upon my back. With the help of the colored woman, the astonished witness of my fall, I finally succeeded in getting upon the horse. Once seated, however, I felt like another person. The vigorous application of a whip, heartily repeated for a few strokes, would arouse the pony into a sullen canter, out of which he would drop with a demonstrative suddenness that made it difficult to keep my seat. In this way considerable relief was obtained for several days from the exasperations produced by the long continuance of pain. After about a fortnight's use of the animal, and when I had learned to be content with half a dozen grains of opium daily, I found myself too weak and helpless to venture on his back, and thus our acquaintance terminated. As this is the first, and probably the last appearance of my equine friend in print, I may as well say that he was sold a short time afterward in the Fifth Street Horse Market, for the sum of forty-three dollars. This is but a meagre price, but the horse had not then become historical.

For the week I was dropping from sixteen grains to nine the addition of new symptoms was slight, but the aggravation of the pain previously endured was marked. The feeling of bodily and mental wretchedness was perpetual, while the tedium of life and occasional vague wishes that it might somehow come to an end were not infrequent. The chief difficulty was to while away the hours of day-light. My rest at night had indeed become imperfect and broken, but still it was a kind of sleep for several hours, though neither very refreshing nor very sound. Those who were about me say that I was in constant motion, but of this I was unconscious. I only recollect that wakening was a welcome relief from the troubled activity of my thoughts. After my morning's ride I usually walked slowly and hesitatingly to the city, but as this occupied only an hour the remaining time hung wearily upon my hands. I could not read—I could hardly sit for five consecutive minutes. Many suffering hours I passed daily either in a large public library or in the book-stores of the city, listlessly turning over the leaves of a book and occasionally reading a few lines, but too impatient to finish a page, and rarely apprehending what I was reading. The entire mental energies seemed to be exhausted in the one consideration—how not to give in to the tumult of pain from which I was suffering. Up to this time I had from boyhood made a free use of tobacco. The struggle with opium in which I was now so seriously engaged had repeatedly suggested the propriety of including the former also in the contest. While the severity of the struggle would, I supposed, be enhanced, the self-respect and self-reliance, the opposition and even obduracy of the will would, I hoped, be enough increased as not seriously to hazard the one great object of leaving off opium forever. Still I dreaded the experiment of adding a feather's weight to the sufferings I was then enduring. An accidental circumstance, however, determined me upon making the trial ; but to my surprise, no inconvenience certainly, and scarce a consciousness of the deprivation accompanied it. The opium suffering was so overwhelming that any minor want was almost inappreciable. The next day brought me down to nine grains of opium. It was now the sixteenth day of December, and I had still fifteen days remaining before the New Year would, as I had resolved, bring me to the complete relinquishment of the drug. The three days which succeeded the disuse of tobacco caused no apparent intensification of the suffering I had been experiencing. On the fourth day, however, and for the fortnight which succeeded, the agony of pain was inexpressibly dreadful, except for the transient intervals when the effects of the opium were felt.

For a few days I had been driven to the alternative of using brandy or increasing the dose of opium. ·I resorted to the former as the least of the two evils. In the condition I was now in it caused no perceptible exhilaration. It did however deaden pain, and made endurance possible. Especially it helped the weary nights to pass away. At this time an entirely new series of phenomena presented themselves. The alleviation caused by brandy was of short continuance. After a few days' use, sleep for any duration, with or without stimulants, was an impossibility. The sense of exhausting pain was unremitted day and night. The irritability both of mind and body was frightful. A perpetual stretching of the joints followed, as though the body had been upon the rack, while acute pains shot through the limbs, only sufficiently intermitting to give place to a sensation of nerveless helplessness. Impatience of a state of rest seemed now to have become chronic, and the only relief I found was in constant though a very uncertain kind of walking which daily threatened to come to an end from general debility. Each morning I would lounge around the house as long as I could make any pretext for doing so, and then ride to the city, for at this time the mud was too deep to think of walking. Once on the pavements, I would wander around the streets in a weary way for two or three hours, frequently resting in some shop or store wherever I could find a seat, and only anxious to get through another long, never-ending day.

The disuse of tobacco, together with the consequences of the diminished use of opium, had now induced a furious appetite. Dining early at a restaurant of rather a superior character, where bread, crackers, pickles, etc., were kept on the table in much larger quantities than it was supposed possible for one individual to need, my hunger had become so extreme that I consumed not only all for which I had specially called, but usually every thing else upon the table, leaving little for the waiter to remove except empty dishes and his own very apparent astonishment. This, it should be understood, was a surreptitious meal, as my own dinner-hour was four o'clock, at which time I was as ready to do it justice as though innocent of all food since a heavy breakfast. The hours intervening between this first and second dinner it was difficult to pass away. The ability to read even a newspaper paragraph had ceased for a number of days. From habit, indeed, I continued daily to wander into several of the city book-stores and into the public library, but the only use I was able to make of their facilities consisted in sitting, but with frequent change of chairs, and looking listlessly around me. The one prevailing feeling now was to get through, somehow or anyhow, the experiment I was suffering under.

Early in the trial my misgivings as to the result had been frequent ; but after the struggle had become thoroughly an earnest one, a kind of cast-iron determination made me sure of a final triumph. The more the agony of pain seemed intolerable, the more seemed to deepen the certainty of my conviction that I should conquer. I thought at times that I could not survive such wretchedness, but no other alternative for many days presented itself to my mind but that of leaving off opium or dying. I recall, indeed, a momentary exception, but the relaxed resolution lasted only as the lightning-flash lasts, though like the lightning it irradiated for a brilliant instant the tumult that was raging within me. For several days previous to this transient weakness the weather had been heavy and lowering, rain falling irregularly, alternating with a heavy Scottish mist. During one of the last days of this protracted storm my old nervous difficulty returned in redoubled strength. Commencing in the shoulder, with its hot needles it crept over the neck and speedily spread its myriad fingers of fire over the nerves that gird the ear, now drawing their burning threads and now vibrating the tense agony of these filaments of sensation. By a leap it next mastered the nerves that surround the eye, driving its forked lightning through each delicate avenue into the brain itself, and confusing and confounding every power of thought and of will. This is neuralgia—such neuralgia as sometimes drives sober men in the agony of their distress into drunkenness, and good men into blasphemy.

While suffering under a paroxysm of this kind, rendered all the more difficult to endure from the exhausted state of the body—in doubt even, at intervals, whether my mind was still under my own control—an impulse of almost suicidal despair suggested the thought, " Go back to opium ; you can not stand this." The temptation endured but for a moment. " No, I have suffered too much, and I can not go back. I had rather die ;" and from that moment the possibility of resuming the habit passed from my mind forever.

It was at night, however, that the suffering from this change of habit became most unendurable. While the day-light lasted it was possible to go out-of-doors, to sit in the sunlight, to walk, to do something to divert attention from the exhausted and shattered body ; but when darkness fell, and these resources failed, nothing remained except a patient endurance with which to combat the strange torment. The only disposition toward sleep was now limited to the early evening. Double dinners, together with the disuse of tobacco, began at this time to induce a fullness of habit in spite of bodily pain. In addition to this, the liver was seriously affected—which seems to be a concomitant of the rapid disuse of opium—and a tendency to heavy drowsiness resulted, as usually happens when this organ is disordered.   As early as six or seven o'clock an unnatural heaviness would oppress the senses, shutting out the material world, but not serving wholly to extinguish the consciousness of pain, and which commonly lasted for an hour or two.   For no longer period could sleep be induced upon any terms.   During these wretched weeks the moments seemed to prolong themselves into hours, and the hours into almost endless durations of time.   The monotonous sound of the ticking clock often became unendurable.   The calmness of its endlessly-repeated beats was in jarring discord with my own tumultuous sensations.   At times it seemed to utter articulate sounds.   "Ret-ri-bu-tion" I recollect as being a not uncommon burden of its song. As the racked body, and the mind, possibly beginning to be diseased, became intolerant of the odious sound, the motion of the clock was sometimes stopped, but the silence which succeeded was even worse to the disordered imagination than the voices which had preceded it.   With the eyes closed in harmony with the deadly stillness, all created nature seemed annihilated, except my single, suffering self, lying in the midst of a boundless void.   If the eyes were opened, the visible world would return, but peopled with sights and sounds that made the misty vastness less intolerable.   There appeared to be nothing in these sensations at all approaching the phenomena exhibited in delirium tremens.   On the contrary, the mind was always and perfectly aware, except for the instant, of the unreal nature of these deceptions and illusions.

A single case will sufficiently illustrate the nature of some
of these apparitions. In the absence of sleep, and while
engaged as was not unusual at this period in the perpetra-
tion of doggerel verse, the irritation of the stomach became
intolerable. The sensation seemed similar to what I had
read of the final gnawings of hunger in persons dying of star-
vation ; a new vitality appeared to be imparted to the organ,
revealing to the consciousness a capacity for suffering previously unsuspected. In the earlier stages, this feeling, which
did not exhibit itself till somewhat late in the process of leaving off opium, was marked by an insatiable craving for stimulus
of some sort, and a craving which would hardly take denial.
While suffering in this way intolerably on one occasion, and
after having attempted in vain to find some possible alleviation suggested in the pages of De Quincey, which lay near
me, I threw myself back on the bed with the old resolution
to fight it out. Almost immediately an animal like a weasel
in shape, but with the neck of a crane and covered with brilliant plumage, appeared to spring from my breast to the floor.
A venerable Dutch market-woman, of whom I had been in
the habit of purchasing celery, seemed to intervene between
me and the animal, begging me not to look at it, and covering it with her apron. Just as I was about to remonstrate
against her interference, something seemed to give way in
the chest and the violence of the pain suddenly abated.

It may aid the reader to form some adequate notion of
the dreary length to which these nights drew themselves
along, to mention that on one occasion, wearied out and disgusted with such illusions, I resolved neither to look at the
clock nor open my eyes for the next two hours. It then
wanted ten minutes to one ; at ten minutes to three my com-
pact with myself would close. For what seemed thousands
upon thousands of times I listened to the clock's steady

ticking. I heard it repeat with murderous iteration, " Ret-ri-bu-tion," varied occasionally, under some new access of pain, with other utterances. Though ordinarily so little endowed with the poetic gift as never to have attempted to write a line of verse, yet at this time, and for a few days previous, I had experienced a strange development of the rhythmical faculty, and on this particular occasion I made verses, such as they were, with incredible ease and rapidity. I remember being greatly troubled by the necessity for a popular national hymn, and manufactured several with extempore rapidity. Had their merit at all corresponded with the frightful facility with which they were composed, they would have won universal popularity. Unfortunately, the effusions were never written down, and can not, therefore, be added to that immense mass of trash which demonstrates the still possible advent of a true American *Marseillaise*.

With these tasks accomplished, and with a suspicion that the allotted hours must have long expired, I would yet remind myself that I was in a condition to exaggerate the lapse of time ; and then, to give myself every assurance of fidelity to my purpose, I would start off on a new term of endurance. I seemed to myself to have borne the penance for hours, to have made myself a shining example of what a resolute will can do under circumstances the most inauspicious. At length, when certain that the time must have much more than expired, and with no little elation over the happy result of the experiment, I looked up to the clock and found it to be just three minutes past one! Little as the mind had really accomplished, the sense of its activity in these few minutes had been tremendous. Measuring time by the conscious succession of ideas may, if I may say it parenthetically, be no more than the same infirmity of our limited human faculties which just now is leading so many men of science,

consciously or unconsciously, to recognize in Nature co-or-
dinate gods, self-subsisting and independent of the ever-liv-
ing and all-present God.

During the five days in which I was descending from the
use of six grains of opium to two, the indications of the
changes going on in the system were these : The gnawing
sensation in the stomach continued and increased ; the
plethoric feeling was unabated, the pulse slow and heavy,
usually beating about forty-seven or forty-eight pulsations to
the minute ; the blood of the whole system seemed to be
driven to the extremities of the body ; my face had become
greatly flushed ; the fingers were grown to the size of thumbs,
while they, together with the palms of the hands and the
breast, parted with their cuticle in long strips. The lower
extremities had become hard, as through the agency of some
compressed fluid. A prickling sensation over the body, as
if surcharged with electricity, and accompanied with an apparent flow of some hot liquid down the muscles of the arms
and legs, exhibited itself at this time. A constant perspiration of icy coldness along the spine had also become a conspicuous element in this strange aggregation of suffering.
The nails of the fingers were yellow and dead-looking, like
those of a corpse ; a kind of glistening leprous scales formed
over the hands ; a constant tremulousness pervaded the
whole system, while separate small vibrations of the fibres on
the back of the hand were plainly visible to the eye. To
these symptoms should be added a dimness of sight often so
considerable as to prevent the recognition of objects even at
a short distance.

With an experience of which this is only a brief outline,
Christmas Day found me using but two grains of opium.
Seven days still remained to me before I was to be brought
by my pledge to myself to the last use of the drug. For

several days previous to this I had abandoned my bed, through apprehension of falling whenever partial sleep left the tumbling and tossing body exempt from the control of the will, and had betaken myself to a low couch made up before the fire, with a second bed on the floor by its side. The necessity for such precaution was repeatedly indicated, but through the kindest care of those whose solicitude never ceased, and who added inexpressibly to this kindness by controlling as far as possible every appearance of solicitude, no injury resulted.

Under the accumulated agony of this part of the trial I began to fear that my mind might give way. I was conscious of occasional fury of temper under very slight provocation. An expressman had charged me what was really an extortionate sum for bringing out a carriage from the city. I can laugh now over the absurd way in which I attacked him, not so much I am sure to save the overcharge as to get rid on so legitimate an object of my accumulated irritability. After nearly an hour's angry dispute, in which I watched successfully and with a malicious ingenuity for any opening through which I could enrage him, and for doing which I am certain he would forgive me if he had known how much I was suffering, he at last gave up the contest by exclaiming, " For heaven's sake give me any thing you please—only let me go!" I had not only saved my money, but felt myself greatly refreshed at finding there was so much life left in me.

It should have been stated before, that when the daily allowance had been reduced to six grains that quantity was divided into twelve pills, and that as this was diminished the size of the pills became gradually smaller till each of them only represented an eighth of a grain. As the daily amount of opium became smaller, although its general effect on the system was necessarily diminished, the conscious relief obtained from each of its fractional parts was for a few minutes more apparent than when these sub-divisions were first made. In this way it was possible so to time the effect as to throw their brief anodyne relief upon the dinner-hour or any other time when it might be convenient to have the agony of the struggle a little alleviated.

While I am not desirous of going into needless detail respecting all the particular phenomena of the process through which I was now passing, it may yet give the reader a more definite idea of the extremely nervous state to which I was reduced, if I mention that so nearly incapable had my hand become of holding a pen, that whenever it was absolutely necessary for me to write a few lines I could only manage it by taking the pen in one quivering hand, then grasp it with the other to give it a little steadiness, watching for an interval in the nervous twitching of the arm and hand, and then, making an uncertain dash at the paper, scrawl a word or two at long intervals. In this way I continued for several weeks to prepare the few brief notes I was obliged to write. My signature at this period I regard with some·curiosity and more pride. It is certainly better than that of Guido Faux, affixed to his examination after torture, though it is hardly equal to the signature of Stephen Hopkins to the Declaration of Independence.

Christmas Day found me in a deplorable condition. No symptom of dissolving nature seemed alleviated; indeed the aggravation of the previous ones, especially of the already unendurable irritation of the stomach, was very obvious. In addition to this, the protracted wakefulness at night began to tell upon the brain, and I resolved to make my case known to a physician. I should have done this long before, but I had been deterred by two things—a long-settled conviction that all recovery from such habits must be essentially the patient's own resolute act, and my misfortune in never having found among my medical friends any one who had made the opium disease a special study, or who knew very much about it. The weather was excessively disagreeable, the heavens, about forty feet off, distilling the finest and most penetrating kind of moisture, while the limestone soil under the influence of the long rain had made walking almost impossible. With frantic impatience I waited until an omnibus made its appearance long after it was due, but crowded outside and in. The only unoccupied spot was the step of the carriage. How in my enfeebled condition I could hold on to this jolting standing-place for half an hour was a mystery I could not divine. With many misgivings I mounted the step, and by rousing all my energies contrived for a few minutes to retain my foot-hold. My knees seemed repeatedly ready to give way beneath me, my sight became dim, and my brain was in a whirl ; but I still held on. I would gladly have left the omnibus, but I was certain that I should fall if I removed my hands from the frame-work of the door by which I was holding on. At length, a middle-aged Irish woman who had been observing me said, "You look very pale, Sir ; I am afraid you are sick. You must take my seat." I thanked her, but told her I feared I had not strength enough to step inside. Two men helped me in, and a few minutes afterward an humble woman was kneeling in her wet clothing in the Church of St.——, not the less penetrated, I trust, with the divine spirit of that commemorative day by her self-denying kindness to a stranger in his extremity. When the paved sidewalk was at last reached I started, after a few minutes' rest, in search of a physician. Purposely selecting the least-frequented streets, in dread of falling if obliged to turn from a direct course, as might be necessary in a crowded thoroughfare, I walked down to the office of the medical man whom I wished to consult; but when I arrived it seemed to me that my case was beyond human aid, and I walked on. I can, perhaps, find no better place than this in which to call the distinct attention of opium-eaters who may be induced to start out on their own reformation, to the all-important fact that no part of the body will be found so little affected by the rapid disuse of opium as the muscles used in walking. I am no physiologist, and do not pretend to explain it, but it is a most fortunate circumstance that in the general chaos and disorder of the rest of the system, the ability to walk, on which so much of the possibility of recovery rests, is by far the least affected of all the physical powers.

During the morning, however, my wretchedness drove me again to the office of the same physician. He listened courteously to my statement; said it was a very serious case, but outside of any reliable observation of his own, and recommended me to consult a physician of eminence residing in quite a different part of the city. He also expressed the hope, though I thought in no very confident tone, that I might be successful, and pretending to shut the door, watched my receding footsteps till I turned a distant corner. I now pass the house of the other physician to whom I was recommended to apply, several times every week, and I often moralize over the apprehension and anxieties with which I then viewed the two or three steps which led to his dwelling. When I arrived opposite his house I stopped and calculated the chances of mounting these steps without falling. I first rested my hand upon the wall and then endeavored to lift my feet upon the second step, but I had not the strength for such an exertion. I thought of crawling to the door, but this was hardly a decorous exhibition for the most fashionable street of the city, filled just then with gayly-dressed ladies. Why I did not ask some gentleman to aid me I can not now recall. I only recollect waiting for several minutes in blank dismay over the seeming impossibility of ever entering the door before me. Finally I went to the curbstone and walked as rapidly and steadily as possible to the lower step, and summoning all my energies made a plunge upward and fortunately caught the door-knob. The physician was at dinner, which gave me some time to recover myself from the agitation into which I had been thrown. After I had narrated my case with special reference to the suspicion of internal inflammation and its possible effect upon the brain, he assured me that no danger of the kind need be anticipated. He hoped I might succeed in my purpose, but thought it doubtful. An uncle of his own, a clergyman of some reputation, had died in making the effort. However, if I would take care of my own resolution, he would answer for my continued sanity. He prescribed some preparation of valerian and red pepper, I think, which I used for a week with little appreciable benefit. Finding no great relief from this prescription, or from those of other medical men whom for a few days about this time I consulted, and feeling a constant craving for something bitter, I at last prescribed for myself. Passing a store where liquor was sold, my eye accidentally rested upon a placard in the window which read " Stoughton's Bitters." This preparation gave me momentary relief, and the only appreciable relief I found in medicine during the experiment.

The nights now began to bring new apprehensions. A constant dread haunted my mind, in spite of the physician's assurances, that my brain might give way from the excitement under which I labored. I was especially afraid of some sudden paroxysm of mania, under the influence of which I might do myself unpremeditated injury. I never feared any settled purpose of self-injury, but I had become nervously apprehensive of possible wayward and maniacal impulses which might result in acts of violence.

My previous business had frequently detained me in the city till a late hour, sometimes as late as midnight. A part of the road that led to my house was quite solitary, with here and there a dwelling or store of the lowest kind. A railroad in process of construction had drawn to particular points on the road small collections of hovels, many of which were whisky-shops, and past these noisy drinking-places it was considered hazardous to walk alone at a late hour. In consequence of the bad reputation of this neighborhood I had purchased a large pistol which I kept ready for an emergency. Now, however, this pistol began to rest heavily upon my mind. The situation of my house was peculiarly favorable for the designs of any marauder. Directly back of it a solitary ravine extended for half a mile or more until it opened upon a populous suburb of the city. This suburb was largely occupied by persons engaged in navigation, or connected with boat-building, or by day-laborers, representing among them many nationalities. The winter of which I am writing was one of unusual stagnation in business and a hard one for the poor to get over. In the nervously susceptible state of my mind at this time, this ravine became a serious discomfort. When the stillness of night settled within and around the house, the rustling of leaves and the distant foot-falls in the ravine became distinctly audible. By some fancy of Judge ——, who built it, the house had no less than seven outside entrances. At intervals I would hear burglars at one of the doors, then at another, nearer or more remote : the prying of levers, the sound of boring, the stealthy footsteps, the carefully-raised window, the heavy breathing of an intruder. Then came the appalling sense of some strange presence, where no outward indication of such presence could be perceived, followed by gliding shadows revealed by the occasional flicker of the waning fire.

Illusions of this nature served to keep the blood at fever-heat during the hours of darkness. Night after night the pistol was placed beneath the pillow in readiness for these ghostly intruders. A few days, however, brought other apprehensions worse than those of thieves and burglars. The uncontrollable exasperation of the temper obliged me at length to draw the charge from the pistol, through fear of yielding to some sudden impulse of despair. I had also put out of reach my razors, a hammer, and whatever else might serve as an impromptu means of violence. I remember the grim satisfaction with which I looked upon the brass ornaments of the bedroom fire-place, and reflected that, if worse came to worst, I was not wholly without a resource with which to end my sufferings. For nearly a fortnight previously I had refrained from shaving, dreading I scarce knew what.

The day succeeding Christmas I rode to the city and walked the length of innumerable by-streets as my weakness would allow. When too exhausted to walk further, and looking for some place of rest, I observed a barber's sign suspended over a basement room. Fortunately the barber stood in the door-way and helped me to descend the half-dozen stone steps which led to his shop. I told the man to cut my hair, shave me, and shampoo my head. As he began his manipulations it seemed as though every separate hair was endowed with an intense vitality. It was impossible to refrain from mingled screams and groans as I repeatedly caught his arm and obliged him to desist. Luckily the barber was a man of sense, and by his extreme gentleness contrived in the course of an hour to calm down my excitement.

When he had finished his work the sense of relief and re-
freshment was astonishing.   In this barber-shop I learned
for the first time in what the perfection of earthly happiness
consists.   The sudden cessation of protracted and severe
pain brings with it so exquisite a sense of enjoyment that I
do not believe that successful ambition, or requited love, or the
gratification of the wildest wishes for wealth, has a happiness
to bestow at all comparable to the calm, contented, all-satisfy-
ing happiness that comes from a remission of intolerable pain.
For the first time in a month I felt an emotion that could be
called positively pleasant.   As I left the shop I needed no
assistance in reaching the sidewalk, and walked the streets
for an hour or two with something of an assured step.

Among other indications of the change taking place at this
time in the system was the increased freezing perspiration
perpetually going on, especially down the spine.   This sense
of dampness and icy coldness has now continued for many
months, and for nearly a year was accompanied with a heavy
cold.   During the opium-eating years I do not remember to
have been affected at all in this latter way ; but a severe cold
at this time settled upon the lungs, one indication of which
was frequent sternutation, consequent apparently upon the
inflammation of the mucous membrane.

In the entire week from Christmas to New Year's the progress in abandonment of opium was but a single grain.   I
am sure there was no want of resolution at this trying time.
Day by day I exhausted all my resources in the vain endeav-
or to get on with half, three-quarters, even seven-eighths of
a grain ; but moans and groans, and biting the tongue till
the blood came, as it repeatedly did, would not carry me
over the twenty-four hours without the full grain.   It seemed
as if tortured nature would collapse under any further effort
to bring the matter to a final issue.

Brandy and bitters after a few day's use had been abandoned, under the apprehension that they were connected with the tendency to internal inflammation which I have noticed as possibly affecting the brain. For a day or two I resorted to ale, but a disagreeable sweetness about it induced the substitution of Schenck beer, a weak kind of *lager*. This I found satisfied the craving for a bitter liquid, and it became for two or three weeks my chief drink. I should have mentioned that the day subsequent to the disuse of tobacco I had also given up tea and coffee, partly from a disposition to test the strength of my resolution, and partly from the belief that they might have some connection with a constant sensation in the mouth as if salivated with mercury. I soon learned that the real difficulty lay in the liver, and that this organ is powerfully affected in persons abandoning the long-continued use of opium. Had I known this fact at an earlier day it would have been of service in teaching me to control the diseased longing for rich and highly-seasoned food which had now become a passion. Eat as much as I would, however, the sense of hunger never left me ; and this diseased craving, in ignorance of its injurious effects, was gratified in a way that might have taxed unimpaired powers of digestion.

At length the long-anticipated New Year's Day, on which I was to be emancipated forever from the tyranny of opium, arrived. For five weeks of such steady suffering as the wealth of all the world would not induce me to encounter a second time, I had kept my eye steadily fixed upon this day as the beginning of a new life. This was also the day on which I was to dine with my friend. As the dinner-hour approached it became evident that no opium meant no dinner, and a little later, that dinner or no dinner the opium was still a necessity. A half grain I thought might carry me through the day, but in this I was mistaken. As I lay upon my friend's sofa, suffering from a strange medley of hunger, pain, and weakness, it seemed that years must elapse before the system could regain its tone or the bodily sensations become at all endurable. Soon after dinner I felt obliged to take another half-grain. My humiliation in failing to triumph when and how I had resolved to do, was excessive. In spite of the strongest resolutions, I was still an opium-eater. I somehow felt that after all I had gone through I ought to have succeeded. I was in no mood to speculate about the causes of the failure; it was enough to know that I had failed, and what was worse, that apparently nothing whatever had been gained in the last four days. While I certainly felt no temptation to give in, I thought it possible that some of the functions of the body, from the long use of opium, might have completely lost their powers of normal action, and that I should be obliged to continue a very moderate use of the drug during the remainder of my life. I saw, in dismal perspective, that small fractional part of the opium of years which was now represented by a single grain, looming up in endless distance, not unlike that puzzling metaphysical necessity in the perpetual subdivision of a unit, which, carried as far as it may be, always leaves a final half undisposed of. But in this I did myself injustice. I had really gained much in these few days, and the proof of it lay in the use of but half a grain on the day which succeeded New Year's. The third day of January, greatly to my surprise, a quarter-grain I found carried me through the twenty-four hours with apparently some slight remission of suffering.

As I now look back upon it, the worst of the experiment lay in the three weeks intervening between the 10th and the 31st of December. So far as mere pain of body was concerned, there was little to choose between the agony of one day and another; but the apprehension that insanity might set in, certainly aggravated the distress of the later stages of the trial. When a man knows that he is practicing self-control to the very utmost, and holding. himself up steadily to his work in spite of the gravest discouragements, the consciousness that a large vacuum is being gradually formed in his brain is not exhilarating.

The next day—to me a very memorable one—the fourth of January, I sat for most of the day rocking backward and forward on a sofa or a chair, speaking occasionally a few words in a low sepulchral voice, but with the one bitter feeling, penetrating my whole nature, that come what would, on that day *I would not.*

When the clock struck twelve at midnight, and I knew that for the first time in many years I had lived for an entire day without opium, it excited no surprise or exultation. The capacity for an emotion of any kind was exhausted. I seemed as little capable of a sentiment as a man well could be, this side of his winding-sheet. I knew, of course, that in these forty days save one, I had worked out the problem, How to leave off opium, and that I had apparently attained a final deliverance : but it was several weeks before I appreciated with any confidence the completion of the task I had undertaken.

Although the opium habit was broken, it was only to leave me in a condition of much feebleness and suffering. I could not sleep, I could not sit quietly, I could not lie in any one posture for many minutes together. The nervous system was thoroughly deranged. Weak as I had become, I felt a continual desire to walk. The weather was unfavorable, but I managed to get several miles of exercise almost daily. But this relief was limited to four or five hours at most, and left the remainder of the day a weary weight upon my hands. The aversion to reading had become such that some months elapsed before I took up a book with any pleasure. Even the daily papers were more than I could well fix my attention upon, except in the briefest and most cursory way. Within a week, however, the sense of acute pain rapidly diminished, but the irritability, impatience, and incapacity to do any thing long remained unrelieved. The disordered liver became apparently more disordered with the progress of time, producing such effects upon the bowels as may with more fitness be told a physician than recorded here. The tonsils of the throat were swollen, the throat itself inflamed, while the chest was penetrated with what seemed like pulsations of prickly heat. There was also a sense of fullness in the muscles of the arms and legs which seemed to be permeated, if I may so express it, with heated electricity. The general condition of the nervous system will be sufficiently indicated by the statement that it was between three or four months before I could hold a pen with any degree of steadiness. Meantime, singular as it may seem, the appearance of health and vigor had astonishingly increased. I had gained more than twenty pounds in weight, partly, I suppose, the result of leaving off opium and tobacco, and partly the consequence of the insatiable appetite with which I was constantly followed. Within a month after the close of the opium strife, I was repeatedly congratulated upon my healthy, vigorous condition. Few men in the entire city bore about them more of the appearance of perfect health, and fewer still were probably in such a state of exhausted vitality.

During the time I was leaving off opium I had labored under the impression that the habit once mastered, a speedy restoration to health would follow. I was by no means prepared, therefore, for the almost inappreciable gain in the weeks which succeeded, and in some anxiety consulted a number of physicians, who each suggested in a timid way the trial, some of strychnine, some of valerian, some of lupuline, hyoscyamus, ignatia, belladonna, and what not. I do not know that I derived the slightest benefit from any of these prescriptions, or from any other therapeutic agency, unless I except the good effects for a few days of bitters, and of cold shower-baths from a tank in which ice was floating.

The most judicious of the medical gentlemen whose aid I invoked, was, I think, the one who replied to my inquiry for his bill, "What for? I have done you no good, and have learned more from you than you have from me."

This constitutes the entire history of my medical experience, and is mentioned as being the only, and a very small adjunct to the great remedy—patient, persistent, obstinate endurance. So exceeding slow has been the process toward the restoration of a natural condition of the system, that writing now, at the expiration of more than a year since opium was finally abandoned, it seems to me very uncertain when, if ever, this result will be reached. Between four and five months elapsed before I was at all capable of commanding my attention or controlling the nervous impatience of mind and body. I then assented to a proposal which involved the necessity of a good deal of steady work, in the hope that constant occupation would divert the attention from the nervousness under which I suffered and would restore the self-reliance which had so long failed me. It was a foolish experiment, and might have proved a fatal one. The business I had undertaken required a clear head and average health, and I had neither. The sleep was short and imperfect, rarely exceeding two or three hours. The chest was in a constant heat and very sore, while the previous bilious difficulties seemed in no way overcome. The mouth was parched, the tongue swollen, and a low fever seemed to have taken entire possession of the system, with special and peculiar exasperations in the muscles of the arms and legs.

The difficulty of thinking to any purpose was only equalled by the reluctance with which I could bring myself to the task of holding a pen. For a few weeks, however, the necessity of not wholly disgracing myself forced me on after a poor fashion ; but at the end of two months I was a used-up man. I would sit for hours looking listlessly upon a sheet of paper, helpless of originating an idea upon the commonest of subjects, and with a prevailing sensation of owning a large emptiness in the brain, which seemed chiefly filled with a stupid wonder when all this would end.

More than an entire year has now passed, in which I have done little else than to put the preceding details into shape from brief memoranda made at the time of the experiment. While the physical agony ceased almost immediately after the opium was abandoned, the irritation of the system still continues. I do not know how better to describe my present state than by the use of language which professional men may regard as neither scientific nor accurate, but which will express, I hope, to unprofessional readers the idea I wish to convey, when I say that the entire system seems to me not merely to have been poisoned, but saturated with poison. Had some virus been transfused into the blood, which carried with it to every nerve of sensation a sense of painful, exasperating unnaturalness, the feeling would not, I imagine, be unlike what I am endeavoring to indicate.

ADDENDA.—At the time of writing the preceding narrative I had supposed that the entire story was told, and that the intelligent reader, should this record ever see the light, would naturally infer, as I myself imagined would be the case, that the unnatural condition of the body would soon become changed into a state of average health. In this I was mistaken. So tenacious and obstinate in its hold upon its victim is the opium disease, that even after the lapse of ten years its poisonous agency is still felt. Without some reference to these remoter consequences of the hasty abandonment of confirmed habits of opium-eating, the chief object of this narrative as a guide to others (who will certainly need all the information on the subject that can be given them) would fail of being secured. While unquestionably the heaviest part of the suffering resulting from such a change of habit belongs to the few weeks in which the patient is abandoning opium, it ought not to be concealed that this brief period by no means comprises the limit within which he will find himself obliged to maintain the most rigid watch over himself, lest the feeling of desperation which at times assaults him from the hope of immediate physical restoration disappointed and indefinitely postponed, should drive him back to his old habits. Indeed, with some temperaments, the greatest danger of a relapse comes in, not during the process of abandonment, but after the habit has been broken. Great bodily pain serves only to rouse up some natures to a more earnest strife, and, as their sufferings become more intense, the determination not to yield gains an unnatural strength. The mind is vindicating itself as the master of the body. While in this state, tortures and the fagot are powerless to extort groans or confessions from the racked or half-consumed martyr. Many a sufferer has borne the agony of the boots or the thumb-screw without flinching, whose courage has given way under the less painful but more unendurable punishment of prolonged imprisonment. In the one case all a man's powers of resistance are roused ; he feels that his manhood is at stake, and he endures as men will endure when they see that the question how far

C

they are their own masters, is at issue. There are, I think, a great number of men and women who would go unflinchingly to the stake in vindication of a principle, whose resolution, somewhere in the course of a long, solitary, and indefinite imprisonment, would break down into a discreditable compromise of opinions for which they were unquestionably willing to die.

In the same way a man will for a time endure even frightful suffering in relinquishing a pernicious habit, while he may fail to hold up his determination against the assaults of the apparently never-ending irritation, discomfort, pain, and sleeplessness which may be counted on as being, sometimes at least, among the remoter consequences of the struggle in which he has engaged. I wish it, however, distinctly understood that I do not suppose that the experience of others whose use of opium had been similar to my own, would necessarily correspond to mine in all or even in many respects. Opium is the Proteus of medicine, and science has not yet succeeded in tearing away the many masks it wears, nor in tracing the marvellously diversified aspects it is capable of assuming. Among many cases of the relinquishment of opium with which I have been made acquainted, nothing is more perplexing than the difference of the specific consequences, as they are exhibited in persons of different temperaments and habits. For such differences I do not pretend to account. That is the business of the thoroughly educated physician, and no unprofessional man, however wide his personal experience, has the right to dogmatize or even to express with much confidence settled opinions upon the subject. My object will be fully attained if I succeed in giving a just and truthful impression of the more marked final consequences of the hasty disuse of opium in this single case, leaving it to medical men to explain the complicated relations of an opium-saturated constitution to the free and healthy functions of life.

In my own case, the most marked among the later consequences of the disuse of opium, some of which remain to the present time and seem to be permanently engrafted upon the constitution, have been these :

1. Pressure upon the muscles of the limbs and in the extremities, sometimes as of electricity apparently accumulated there under a strong mechanical force.

2. A disordered condition of the liver, exhibiting itself in the variety of uncomfortable modes in which that organ, when acting irregularly, is accustomed to assert its grievances.

3. A sensitive condition of the stomach, rejecting many kinds of food which are regarded by medical men as simple and easy of digestion.

4. Acute shooting pains, confined to no one part of the body.

5. An unnatural sensitiveness to cold.

6. Frequent cold perspiration in parts of the body.

7. A tendency to impatience and irritability of temper, with paroxysms of excitement wholly foreign to the natural disposition.

8. Deficiency and irregularity of sleep.

9. Occasional prostration of strength.

10. Inaptitude for steady exertion.

I mention without hesitancy these consequences of the abandonment of opium, from the belief that any person really in earnest in his desire to relinquish the habit will be more likely to persevere by knowing at the start exactly what obstacles he may meet in his progress toward perfect recovery, than by having it gradually revealed to him, and that at times when his body and mind are both enfeebled by what he has passed through. With a single exception, the disturbance from the first of these causes has been much the most serious one I have been obliged to encounter. Whether it is one of the specific effects of the disuse of opium, or only one of the many general results of a disordered constitution, I do not know.

I can only say in my own case, that after the lapse of years, this particular difficulty is not wholly overcome. This electric condition, so to call it, still continues a serious annoyance. But when it occurs, the pain is of less duration, and gradually, but very slowly, is of diminished frequency. Violent exercise will sometimes relieve it ; a long walk has often the same effect. The use of stimulants brings alleviation for a time, but there seems to be no permanent remedy except in the perfect restoration of the system by time from this effect of the wear and tear of opium upon the nerves. Irregularity in the action of the liver, while singularly marked in the earlier stages of the experiment, and continuing for years to make its agency manifestly felt, is in a considerable degree checked and controlled by a judicious use of calomel.

The condition of the digestive organs is less impaired than I should have supposed possible, judging from the experience of others. A moderate degree of attention to the quality of what is eaten, with proper care to avoid what is not easily digested, with the exercise of habitual self-control in respect to quantity, suffices to prevent, for the most part, all unendurable feelings of discomfort in this part of the system. Whether the habitually febrile condition of the mouth, and the swollen state of the tongue, is referable to a disturbed action of the stomach or of the liver I can not say. It is certain that none of the effects of opium-eating are more marked or more obstinately tenacious in their hold upon the system than these. I barely advert to the frequent impossibility of retaining some kinds of food upon the stomach,

which has been one unpleasant part of my experience, because I doubt whether this return of a difficulty which began
in childhood has any necessary connection with the use of
opium. For many years before I knew any thing of the
drug I had been a daily sufferer from this cause. Indeed
the use of opium seemed to control this tendency, and it was
only when the remedy was abandoned that the old annoyance returned. For a few months the stomach rejected every
kind of food ; but in less than a year, and subsequently to
the present time, this has been of only occasional occurrence.

I am also at a loss how far to connect the disuse of opium with the lancinating pains which have troubled me
since the time to which I refer. These pains began long
before I had recourse to opium, they did not cease their frequent attacks while opium was used, nor have they failed to
make their potency felt since opium was abandoned. While
it is not improbable that the neuralgic difficulties of my
childhood might have remained to the present time, even if
I had never made use of opium, I think that the experience
of all who have undergone the trial shows that similar
pains are invariably attendant upon the disuse of opium.
How long their presence might be protracted with persons
not antecedently troubled in this way, is a question I can
not answer. I infer from what little has been recorded, and
from what I have learned in other ways, that the reforming
opium-eater must make up his mind to a protracted encounter with this great enemy to his peace. That the struggle of
others with this difficulty will be prolonged as mine has been
I do not believe, unless they have been subjected for a lifetime to pains connected with disorder in the nervous system.

The unnatural sensitiveness to cold to which I have alluded is rather a discomfort than any thing else. It merely
makes a higher temperature necessary for enjoyment, but in

no other respect can it be regarded as deserving special mention. With the thermometer standing at 80° to 85° the sensation of agreeable warmth is perfect ; with the mercury at 70° or even higher, there is a good deal of the feeling that the bones are inadequately protected by the flesh, that the clothing is too limited in quantity, and in winter that the coal-dealer is hardly doing you justice.

The cold perspiration down the spine, which was so marked a sensation during the worst of the trial, has not yet wholly left the system, but is greatly limited in the extent of surface it affects and in the frequency of its return.

The tendency to impatience and irritability of temper to which I have adverted is by far the most humiliating of the effects resulting from the abandonment of opium. Men differ very widely both in their liability to these excesses of temper as well as in their power to control them ; but under the aggravations which necessarily attend an entire change of habit, this natural tendency, whether it be small or great, to hastiness of mind is greatly increased. So long as the disturbing causes remain, whether these be the state of the liver or the stomach, or a want of sufficient sleep, or the excited condition of the nervous system, the patient will find himself called upon for the exercise of all his self-control to keep in check his exaggerated sensibility to the daily annoyances of life.

Intimately connected with the preceding is the frequent recurrence of sleepless nights, which seem invariably to attend upon the abandonment of the habit. Possibly some part of this state of agitated wakefulness may pertain to the natural temperament of the patient, but this tendency is greatly aggravated by the condition of the nerves, so thoroughly shattered by the violent struggle to oblige the system to dispense with the soothing influence of the drug upon which it has so long relied.   Whatever method others may have found to counteract this infirmity, I have been able as yet to find no remedy for it.   Especially are those nights made long and weary which *precede* any long continuance of wet weather.   A moist condition of the atmosphere still serves the double purpose of setting in play the nervous sensibilities, and, as a concomitant or a consequence, of greatly disturbing, if not destroying sleep.

In connection with this matter something should be said on the subject of dreaming, to which De Quincey has given so marked a prominence in his " Confessions " and " Suspiris de Profundis."   In my own case, neither when beginning the use of opium, nor while making use of it in the largest quantities and after the habit had long been established, nor while engaged in the painful process of relinquishing it, nor at any time subsequently, have I had any experience worth narrating of the influence of the drug over the dreaming faculty.   On the contrary, I doubt whether many men of mature age know so little of this peculiar state of mind as myself. The conditions in this respect, imposed by my own peculiarities of constitution, have been either no sleep sufficiently sound as to interfere with the consciousness of what was passing, or mere restlessness, or sleep so profound as to leave behind it no trace of the mind's activity.   While it is therefore certain that this exaggeration of the dreaming faculty is not necessarily connected with the use of opium, but is rather to be referred to some peculiarity of temperament or organization in De Quincey himself, I find myself in turn at a loss to know how far to regard other phenomena to which I have previously alluded as the natural and necessary consequences of opium, or how far they may be owing to peculiarities of constitution in myself.   Opium-eaters have said but little on the subject.   The medical profession, so far as I have conversed with them, and I have consulted with some of the most eminent, are not generally well informed on any thing beyond the specific effects of the drug as witnessed in ordinary medication.   In the absence of sufficient authority, it may be safer to say that the remoter consequences of the disuse of opium consist in a general disorder and derangement of the nervous system, exhibiting itself in such particular symptoms as are most accordant with the temperament, constitutional weaknesses, and personal idiosyncrasies of the patient.   That some considerable suffering must be regarded as unavoidable seems to be placed beyond question from the nature of the trial to which the body has been subjected, as well as from what little has been said on the subject by those who have relinquished the habit.

I close this brief reference to the remoter consequences of the habits of the opium-eater by calling the attention of the reader to the physical weakness with consequent inaptitude for continuous exertion which forms a part of my own experience.   Unable as I am to refer it to any *immediate* cause, frequent and sudden prostration of strength occurs, accompanied by slight dizziness, impaired sight, and a sense of overwhelming weakness, though never going to the extent of absolute fainness.   Its recurrence seems to be governed by no rule.   It sometimes comes with great frequency, and sometimes weeks will elapse without a return.   Neither the state of the weather, nor any particular condition of the body, appears to call it out.   It sometimes is relieved by a glass of water, by the entrance of a stranger, by the very slightest excitement, and it sometimes resists the strongest stimulants and every other attempt to combat it.   I can record nothing else respecting this visitant except that its presence is always accompanied with a singular sensation in the stomach, and that the entire nervous system is affected by its attack.

The inaptitude for steady exertion is not merely the consequence of this occasional feeling of exhaustion, but is for a time the inevitable result of the accumulated pain and weakness to which his system, not yet restored to health, is still subject. This impatience of continued application to work, which is common to all opium-eaters, and which does not cease with the abandonment of the habit, seems to result in the first case from some specific relation between the drug and the meditative faculties, promoting a state of habitual reverie and day-dreaming, utterly indisposing the opium-user for any occupation which will disturb the calm current of his thoughts, and in the other, proceeding from the direct disorder of the nervous organization itself. Strange as it may seem, the very thought of exertion will often waken in the reforming opium-eater acute nervous pains, which cease only as the purpose is abandoned. In other cases, where there is no special nervous suffering at the time, work is easy and pleasant even beyond what is natural.

One effect of opium upon the *mind* deserves to be mentioned ; its influence upon the faculty of memory. The logical memory, De Quincey says, seems in no way to be weakened by its use, but rather the contrary. His own devotion to the abstract principles of political economy ; the character of Coleridge's literary labors between the years 1804–16, when his use of opium was most inordinate ; together with the cast of mind of many other well-known opium-eaters, confirms this suggestion of De Quincey. His further statement that the memory of dates, isolated events, and particular facts, is greatly weakened by opium, is confirmed by my own experience. However physiologists may explain this fact, a knowledge of it may not be without its use to those who desire to be made thoroughly acquainted with all the consequences of the opium habit.

If to these discomforts be added a prevailing tendency to a febrile condition of body, together with permanent disorder in portions of the secretory system, the catalogue of annoyances with which the long-reformed opium-eater may have to contend is completed. This statement is not made to exaggerate the suffering consequent upon the disuse of opium, but is made on the ground that a full apprehension of what the patient may be called upon to go through will best enable him to make up his mind to one resolute, unflinching effort for the redemption of himself from his bad habits.

So far as the body is concerned, there is much in my experience which induces me to give a general assent to the opinion expressed by a medical man of great reputation whom I repeatedly consulted in reference to the discouraging slowness of my own restoration to perfect health. "I can not see," he said, "that your constitution has been permanently injured ; but you were a great many years getting into this state, and I think it will take nearly as many to get you out of it."

It may not be amiss to add that those opium-eaters whose circumstances exempt them from harassing cares, who meet only with kindness and sympathy from friends, and who have resources for enjoyment within themselves, have in respect to these subsequent inconveniences greatly the advantage of those whose position and circumstances are less fortunate.

These free and almost confidential personal statements have been made, not without doing some violence to that instinctive sense of propriety which prompts men to shrink from giving publicity to their weaknesses and from the vanity of seeming to imply that their individual experience of life is of special value to others. Leaving undecided the question whether under any circumstances a departure from the general rule of good sense and good taste in such mat-

ters is justifiable, I have, nevertheless, done what I could
to give to opium-eaters a truthful statement of the consequences that may ensue from their abandonment of the habit. The path toward perfect recovery is certainly a weary
one to travel ; but in all these long years, with nervous sensibilities unnaturally active, in much pain of body, through innumerable sleepless nights, with hope deferred and the expectation of complete restoration indefinitely prolonged, I
have never lost faith in the final triumph of a patient and
persistent resolution. Many men seem to know little of
the wonderful power which simple endurance has, in deter-
mining every conflict between good and evil. The triumph
which is achieved in a single day is a triumph hardly worth
the having ; but when all impatience, unreasonableness,
weaknesses and vanities have been burned out of our natures
by the heat of suffering ; when the resolution never falters
to endure patiently whatever may come in the endeavor to
measure one's own case justly, and exactly as it is ; and when
time has been allowed to exert its legitimate influence in
calming whatever has been disturbed· and correcting whatever has been prejudiced, a conscious strength is developed
far beyond what is natural to men possessed only of ordinary powers of endurance. It is chiefly through patient
waiting that the confirmed victim of opium can look for
relief. All who have made heroic efforts to this end, and
yet have failed in their attempt, have done so through the
absence of adequate confidence in the efficacy of time to
bring them relief. The *one* lesson, however, which the reforming opium-eater must learn is, never to relinquish any
gain, however slight, which he may make upon his bad habit.
Patience will bring him relief at last, and though he may and
will find his progress continually thwarted and himself often
tempted to give over the contest in despair, he may be sure

that year by year he is steadily advancing to the perfect recovery of all that he has lost.

The opium-eater will not regard as amiss some few suggestions as to the mode in which his habit may most easily be abandoned. The best advice that can be given—the *only* advice that will ever be given by an opium-eater—is, never to begin the habit. The objection at once occurs, both to the medical man and to the patient suffering from extreme nervous disorder, What remedy then shall be given in those numerous cases in which the protracted use of opium, laudanum, or morphine is found necessary? The obvious answer is, that no medical man ever intends to give this drug in such quantities or for so long a time as to establish in the patient a confirmed habit. The frequent, if not the usual history of confirmed opium-eaters is this: A physician prescribes opium as an anodyne, and the patient finds from its use the relief which was anticipated. Very frequently he finds not merely that his pain has been relieved, but that with this relief has been associated a feeling of positive, perhaps of extreme enjoyment. A recurrence of the same pain infallibly suggests a recurrence to the same remedy. The advice of the medical man is not invoked, because the patient knows that morphine or laudanum was the simple remedy that proved so efficacious before, and this he can procure as well without as with the direction of his physician. He becomes his own doctor, prescribes the same remedy the medical man has prescribed, and charges nothing for his advice. The resort to this pleasant medication after no long time becomes habitual, and the patient finds that the remedy, whose use he had supposed was sanctioned by his physician, has become his tyrant. If patients exhibited the same reluctance to the administration of opium that they do to drugs that are nauseous, if the collateral effects of the former were no more pleasurable than lobelia or castor oil, nothing more could be said against self-medication in one case than the other. Opium-eaters are made such, not by the physician's prescription of opium to patients in whose cases its use is indispensable, but by their not giving together with such prescriptions emphatic and earnest caution that the remedy is not to be taken except when specially ordered, in consequence of the hazard that a habit may be formed which it will be difficult to break. Patients to whom it is regularly administered are not at first generally aware how easily this habit is acquired, nor with what difficulty it is relinquished, especially by persons of nervous temperament and enfeebled health. The number of cases, I suspect, is small in which the use of opium has become a necessity, where the direction of a physician may not be pleaded as justifying its original employment.

The object I have in view is not, however, so much to make suggestions to medical men as it is to awaken in the victims of opium the feeling that they can master the tyrant by such acts of resolution, patience, and self-control as most men are fully capable of exhibiting. Certain conditions, however, seem to be the almost indispensable preliminaries to success in relinquishing opium by those who have been *long* habituated to its use. The first and most important of these is a firm conviction on the part of the patient that the task can be accomplished. Without this he can do nothing. The narratives given in this volume show its entire practicability. In addition to this, it should be remembered that these experiments were most of them made in the absence of any sufficient guidance, from the experience of others, as to the method and alleviations with which the task can be accomplished. A second condition necessary to success, is sufficient physical health, with sufficient firmness of character to undergo, as a matter of course, the inevitable suffering of the body, and to resist the equally inevitable temptation to the mind to give up the strife under some paroxysm of impatience, or in some moment of dark despondency. With a very moderate share of vigor of constitution, and with a will, capable under other circumstances of strenuous and sustained exertion, there is no occasion to anticipate a failure here. Even in cases of impaired health, and with a diminished capacity for resolute endeavor, success is, I believe, attainable, provided sufficient time be taken for the trial.

A further condition lies in the attempt being made under the most favorable circumstances in respect to absolute leisure from business of every kind. That nothing can be accomplished by persons whose time is not at their own command, by a graduated effort protracted through many months, I do not say, for I do not believe it ; but any speedy relinquishment of opium—that is, within a month or two—seems to me to be wholly impossible, except to those who are so situated that they can give up their whole time and attention to the effort.

This effort should be made with the advice and under the eye of an intelligent physician. So far as I have had opportunity to know, the profession generally is not well informed on the subject. In my own case I certainly found no one who seemed familiar with the phenomena pertaining to the relinquishment of opium, or whose suggestions indicated much more than vague, empirical ideas on the subject. But even in cases where the physician has had no experience whatever in this class of disorders, he can, if a well-educated man, bring his medical knowledge and medical reasoning to bear upon the various states, both of body and mind, which the varying sufferings of the patient may make known to him. Were there, indeed, no professional helps to be secured by such consultation, it is still of infinite service to the patient to know some one to whom he can frequently impart the history of his struggle and the progress he is making. Such confidence may do much to encourage the patient, and no one is so proper a person in whom to repose this confidence as an intelligent physician.

The amount of time which should be devoted to the experiment must depend very greatly upon these considerations —the constitution of the patient, the length of time which has elapsed since the habit was formed, and the quantity habitually taken. When the habit is of recent date, and the daily dose has not been large—say not more than ten or twelve grains—if the patient has average health, his emancipation from the evil may be attained in a comparatively short period, though not without many sharp pangs and many wakeful nights which will call for the exercise of all his resolution.

The question will naturally suggest itself to others, as it has often done to myself, whether a less sudden relinquishment of opium would not be preferable as being attended with less present and less subsequent suffering. Numerous cases have come under my notice where a very gradual reduction was attempted, but which resulted in failure. Only two exceptions are known to me : in one of these the patient, himself a physician, effected his release by a graduated reduction extending through five months. The other is the case of Dr. S., a physician of eminence in Connecticut many years ago. This gentleman had made so free use of opium to counteract a tendency to consumption that the habit became established. After several years, and at the suggestion of his wife, he made a resolution to abandon it, engaging to take no opium except as it passed through her hands, but with the understanding that the process of relinquishment was to be slow and gradual. His allowance at this time was understood to be from twenty to thirty grains of crude opium daily. At the end of two years the habit was abandoned, with no very serious suffering during the time, and so far as his daughter was informed, with no subsequent inconvenience to himself. He lived many years after his disuse of opium, in the active discharge of the duties of his profession, and died at last in the ninetieth year of his age. The hazard of this course, however, consists in the possibility, not to say with some temperaments the probability, that somewhere in the course of so very gradual a descent the same influences which led originally to the use of opium may recur, with no counteracting influence derived from the excitement of the mind produced by the earnestness of the struggle. With some constitutions I have no doubt that a process even so slow as that of Dr. S.'s might be successful, but I suspect, with most men, that some mood of excited feeling, and some conscious sense of conflict, will be found necessary, in order to bring them up resolutely to the work of self-emancipation. On the other hand, I am satisfied that my own descent was too rapid. Had the experiment of between five and six weeks been protracted to twice that time, much of the immediate suffering, and probably more of that which soon followed, might have been prevented. As in the constitution of every person there is a limit beyond which further indulgence in any pernicious habit results in chronic derangement, so also there seems to be a limit in the discontinuance of accustomed indulgence, going beyond which is sure to result in some increased physical disorder. In the cure of *delirium tremens*, the first step of the physician is to stimulate. With more moderate drinkers abrupt cessation from the use of stimulants is the only sure remedy. In the first instance the nervous system is too violently agitated to dispense entirely with the accustomed habit; in the second, the nerves are presumed to be able to bear the temporary strain imposed upon them by the condition of the stomach and other organs. But with opium the case is otherwise. Insanity, I think, would be the general result of an attempt immediately to relinquish the habit by those who have long indulged it. The most the opium-eater can do is to diminish his allowance as rapidly as is safe. For the same reason that no sensible physician would direct the confinement of a patient and the absolute disuse of opium with the certainty that mania would result, so it would be equally ill advised to recommend a diminution so rapid as necessarily to call out the most serious disorder and derangement of all the bodily functions, especially if these could be made more endurable by being spread over a longer period. In one respect the opium-eater has greatly the advantage over those addicted to other bad habits. Those who have used distilled or fermented drinks, tobacco, and sometimes coffee and tea in excess, experience for a time a strong and definite craving for the wonted indulgence. This is never the case with the opium-eater ; he has no specific desire whatever for the drug. The only difficulty he has to encounter is the agony of pain—for no other word adequately expresses the suffering he endures—conjoined with a general desire for relief. Yet in the very *acme* of his punishment he will be sensible of no craving for opium at all like the craving of the drunkard for spirits. As De Quincey justly represents it, the feeling is more that of a person under actual torture, aching for relief, though with no care from what source that relief comes. So far from there being any particular desire for opium, there ensues very speedily, I suspect, after the attempt to abandon it is begun, and long before the necessity for its use has ceased, and even while the suffering from its partial disuse is most unendurable, a feeling in reference to the drug itself not far removed from disgust. The only occasion that I have had of late years to make use of opium or any of its preparations, was within a twelvemonth after it had been laid aside. A morbid feeling had long troubled me with the suggestion that should a necessity ever arise for the medical use of opium, I might be precipitated back into the habit. I was not sorry, therefore, when the necessity for its use occurred, that I might test the correctness of my apprehension. To my surprise, not only was no desire for a second trial of its virtues awakened, but the very effort to swallow the pill was accompanied with a feeling akin to loathing.

The final decision of the question, How long a time should be allowed for the final relinquishment of the drug? must, I imagine, be left to a wider experience than has yet been recorded. The general strength of the constitution, the force of the will, the degree of nervous sensibility, together with the external circumstances of one's life, have all much to do with its proper explication.

The general directions I should be disposed to suggest for the observance of the confirmed opium-eater would be something as follows :

1. To diminish the daily allowance as rapidly as possible to one-half. A fortnight's time should effect this without serious suffering, or any thing more than slight irritation and some other inconveniences that will be found quite endurable to one who is in earnest in his purpose.

2. For the first week, if the previous habit has been to take the daily dose in a single portion, or even in two portions, morning and night, it will be found advisable to divide the diminished quantity into four parts. Thus, if eighty grains has been the customary quantity taken, four pills of fifteen

grains each, taken at regular intervals, say one at eight and one at twelve o'clock in the morning, and one at four and one at eight in the evening, will be found nearly equal in their effect to the eighty grains taken at once in the morning. A further diminution of two grains a day, or of half a grain in each of these four daily portions, will within the week reduce the quantity taken to fifty grains, and this without much difficulty, and with positive gain in respect to elasticity of spirits, arising, in part, from the newly-awakened hope of ultimate success. A second week should suffice for a reduction to forty grains. It will probably be better to divide the slightly diminished daily allowance into five portions, to be taken at intervals of two hours from rising in the morning till the daily quantity is consumed. With such a graduated scale of descent, it will be found at the end of two weeks that one-half of the original quantity of opium has been abandoned, and that, with so little pain of body, and so much gain to the general health and spirits, that the completion of the task will seem to the patient ridiculously easy. He will soon learn, however, that he has not found out all the truth.

⌐ In the third week a further gain of ten grains can the more easily be made by still further dividing the daily portion into an increased number of parts, say ten. The feeling of restlessness and irritability by this time will have become somewhat annoying, and the actual struggle will be seen to have commenced. It will doubtless require at this point some persistence of character to bear up against the increased impatience, both of body and spirit, which marks this stage of the descent. The feelings will endeavor to palm off upon the judgment a variety of reasons why, for a time, a larger quantity should be taken ; but this is merely the effect of the diminished amount of the stimulant. Sleep will probably be found to be of short continuance as well as a good deal broken. Reading has ceased to interest, and a fidgety, fault-finding temper not unlikely has begun to exhibit itself. At this point, I am satisfied, most opium-eaters who have endeavored in vain to renounce the habit, have broken down. Their resolution has failed them not because they were unable to stand much greater punishment than had yet been inflicted, but because they yielded to the impression that some other time would prove more opportune for the final experiment. Under this delusion they have foolishly thrown away the benefit of their past self-control, with the certainty that should the trial be again made, they would once more be assailed by a similar temptation. But if this stage of the process has been safely passed, the next—that of reducing the daily quantity from thirty grains to twenty-five, still dividing the day's allowance into ten portions—would probably have added little aggravation to the uncomfortable feeling which already existed, but not without some conscious addition, on the other hand, to their enjoyment from the partially successful result of the experiment. Thus in four weeks a very substantial gain, by the reduction of the needed quantity from eighty grains to twenty-five, would have been attained.

If the patient should find it necessary to stop at this point for a week, a fortnight, or even longer, no great harm would necessarily result ; it would only postpone by so much his ultimate triumph. He should never forget, however, that the one indispensable condition of success is this : *Never under any circumstances to give up what has been once gained.* If in any manner the patient has been able to get through the day with the use of only twenty-five grains, it is certain that he can get through the next, and the next, and the subsequent day with the same amount, with the further certainty that the habit of being content with this minimum quantity will soon begin to be established, and that speedily a further advance may be made in the direction of an entire disuse. Whenever the patient finds his condition to be somewhat more endurable, whether the time be longer or shorter, he should make a still further reduction, say to one-quarter of his original dose. If this abatement of quantity be spread over the entire week the aggravation of his discomfort will not be great, while the elation of his spirits over what he has already accomplished will go far in enabling him to bear the degree of pain which necessarily pertains to the stage of the experiment which he has now reached. The caution, however, must be borne continually in mind that under no circumstances and on no pretext must the patient entertain the idea that any part of that which he has gained can be surrendered. Better for him to be years in the accomplishment of his deliverance than to recede a step from any advantage he may have secured. If he persists, he will in a few days, or at the longest in a few weeks, find his condition as to bodily pain endurable if nothing more. There may not, probably will not be any very appreciable gain from day to day. The excited sufferer, judging from his feelings alone, may think that he has made no progress whatever; but if after the lapse of a week he will contrast his command of temper, or his ability to fix his attention upon a subject, as evinced at the beginning and end of this period, he can hardly fail to see that there has been a real if not a very marked advance in his status. Such a person has no right to expect, after years of uninterrupted indulgence, that the most obstinate of all habits can be relinquished with ease, or that he can escape the penalty which is wisely and kindly attached to all departures from the natural or supernatural laws which govern the world. It should be enough for him to know that there

is no habit of mind or of body which may not be overcome,
and that the process of overcoming, in its infinite variety of
forms, is that out of which almost all that is good in character
or conduct grows, and that the amount of this good is usually
measured by the struggle which has been found necessary to
ensure success.

Considerations of this nature, however, are of too general
a character to be of much service to one enduring the misery
of the reforming opium-eater.   He has now arrived at a point
where he is obliged to ask himself when and how the contest is to end.   He has succeeded in abandoning three-quarters of the opium to which he has so long been accustomed.
A few weeks have enabled him to accomplish this much.
He endures, indeed, great discomfort by day and by night;
but hope has been re-awakened; his mind has recovered
greater activity than it has known for years; and, on the
whole, he feels that he has been greatly the gainer from the
contest.

Let me repeat, that the main thing for the patient at this
point of his trial is not to forego the advantage he has already attained—"not to go back."   If he can only hold his
own he has so far triumphed, and it is only a question of
time when the triumph shall be made complete.   *When* this
shall be effected *he* must decide.   The rapidity of his further
progress must be determined by what he himself is conscious
he has the strength, physical and moral, to endure.   With
some natures any very sudden descent is impossible; with
others, whatever is done must be done continuously and rapidly or is not done at all.   The one temperament can not
stand up against the assaults of a fierce attack, the other
loses courage except when the fight is at the hottest.   For
the former ample time must be given or he surrenders; the
latter will succumb if any interval is allowed for repose.   It

is, therefore, difficult to suggest from this point downward any rule which shall apply equally to temperaments essentially unlike. I think, however, that the suggestion to divide the daily allowance, whether the descent be a slow or a rapid one, into numerous small parts to be taken at equal intervals of time, will be found to facilitate the success of the attempt in the case of both. The chief value of such subdivision probably consists in its throwing the aggregate influence of the day's opium nearer the hour of bed-time, when it is most needed, than to an earlier hour, when its soporific power is less felt. In addition to this, the importance to the excited and irritated patient of being able to look forward during the long-protracted hours to frequent, even if slight, alleviations of his pain, should not be left out of the account. In general it may be said that whenever the patient feels that he can safely, that is, without danger of failing in his resolution, adventure upon a further diminution of the quantity, an additional amount, smaller or greater according to circumstances, should be deducted till the point is reached where the suffering becomes unendurable ; then after a delay of few or many days, as may be needed to make him somewhat habituated to the diminished allowance, a still further reduction should be made, and so on for such time as the peculiarities of different constitutions and circumstances may make necessary, till the quantity daily required has become so small, say a grain or two, that by still more minute subdivisions, and by dropping one of them daily, the final victory is achieved.

I have not ventured to say in how short a time confirmed habits of opium-eating may be abandoned. In my own case it was thirty-nine days, but with my present experience I should greatly prefer to extend the time to at least sixty days ; and this chiefly with reference to the violent effects upon the constitution produced by the suddenness of the change of habit. Some constitutions may possibly require less time and some probably more. While I regard the abandonment of the first three-quarters of the accustomed allowance as being a much easier task than the last quarter, and one which can be accomplished with comparative impunity in a brief period, I would allow at least twice the time for the experiment of dispensing with the last quarter; unless, indeed, I should be apprehensive that my resolution might break down through the absence of the excitement which is unquestionably afforded by the feeling that you are engaged in a deadly but doubtful conflict. So far, also, as can be inferred from cases subsequently narrated in this volume, the probability of success would seem to be enhanced by devoting a longer time to the trial. It can not, however, be too often repeated, that however slow or however rapid the pace may be, the rule to be rigidly observed is this : Never to increase the minimum dose that has once been attained. This is the only rule of safety, and by adhering to it, persons in infirm health, or with weakened powers of resolution, will ultimately succeed in their efforts.

I subjoin my own record of the quantity of opium daily consumed, for the possible encouragement of such opium-eaters as may be disposed to make trial of their own resources in the endurance of bodily and mental distress.

Saturday,	Nov. 25	80 grains,	= 2000 drops of laudanum.
Sunday,	" 26	60 "	1500 " "
Monday,	" 27	50 "	1250 " "
Tuesday,	" 28	40 "	1000 " "
Wednesday,	" 29	30 "	750 " "
Thursday,	" 30	25 "	625 " "
Friday,	Dec. 1	20 "	500 " "
Average of 1st week		44 "	1089 " "

Saturday,	Dec.	2	19 grains,	=	475	drops of laudanum.	
Sunday,	"	3	18 "		450	"	"
Monday,	"	4	17 "		425	"	"
Tuesday,	"	5	16 "		400	"	"
Wednesday,	"	6	15 "		375	"	"
Thursday,	"	7	15 "		375	"	"
Friday,	"	8	15 "		375	"	"
Average of 2d week .... 16.43 "					411	"	"

Saturday,	Dec.	9	14 grains,	=	350	drops of laudanum.	
Sunday,	"	10	13 "		325	"	"
Monday,	"	11	13 "		325	"	"
Tuesday,	"	12	12 "		300	"	"
Wednesday,	"	13	12 "		300	"	"
Thursday,	"	14	11 "		275	"	"
Friday,	"	15	10 "		250	"	"
Average of 3d week .... 12.14 "					304	"	"

Saturday,	Dec.	16	9 grains,	=	225	drops of laudanum.	
Sunday,	"	17	8 "		200	"	"
Monday,	"	18	8 "		200	"	"
Tuesday,	"	19	7 "		175	"	"
Wednesday,	"	20	6 "		150	"	"
Thursday,	"	21	5 "		125	"	"
Friday,	"	22	4 "		100	"	"
Average of 4th week .... 6.71 "					168	"	"

Saturday,	Dec.	23	3 grains,	=	75	drops of laudanum.	
Sunday,	"	24	3 "		75	"	"
Monday,	"	25	2 "		50	"	"
Tuesday,	"	26	2 "		50	"	"
Wednesday,	"	27	2 "		50	"	"
Thursday,	"	28	2 "		50	"	"
Friday,	"	29	1 "		25	"	"
Average of 5th week .... 2.14 "					54	"	"

Saturday,	Dec.	30	1 grain,	=	25	drops of laudanum.	
Sunday,	"	31	1 "		25	"	"
Monday,	Jan.	1	1 "		25	"	"
Tuesday,	"	2	$\frac{1}{2}$ "		12	"	"
Wednesday,	"	3	$\frac{1}{4}$ "		6	"	"
Average of 6th week .... 0.75 "					18	"	"

D

The fourth and fifth weeks I found to be immeasurably the most difficult to manage. By the sixth week the system had become somewhat accustomed to the denial of the long-used stimulant. At any rate, though no abatement of the previous wretchedness was apparent, it certainly seemed less difficult to endure it. It is at this stage of the process that I regard the advice and encouragement of a physician as most important. He may not indeed be able to do much in direct alleviation of the pain incident to the abandonment of opium, for I suspect that little reliance can be placed upon the medicines ordinarily recommended. The system has become accustomed to the stimulant to an exorbitant degree ; the suffering is consequent upon the effort to accustom the system to get on without it. Other kinds of stimulants, like spirits or wine, will afford a slight relief for a few days, especially if taken in sufficiently large quantities to induce sleep. It is the sedative qualities of the opium that are chiefly missed, for as to excitement the patient has quite as much of it as he can bear. For this reason malt liquors are preferable to distilled spirits—they stupefy more than they excite. But to malt liquors this serious objection exists, they tend powerfully to aggravate all disorders of the liver. This tendency the reforming opium-eater can not afford to overlook, for no one effect of the experiment is more distressing than the marvellous and unhealthy activity given to this organ by the process through which he is passing. The testimony of all opium-eaters on this point is uniform. For months and even years this organ in those who have relinquished the drug remains disordered. When in its worst state, the use of something bitter, the more bitter the better, is exceedingly grateful. The difficulty lies in finding any thing that has a properly bitter taste. Aloes, nux vomica, colocynth, quassia, have a flavor that is much more sweet than bitter. These serious annoyances from the condition of the liver, as well as those arising from the state of the stomach and some of the other organs, may be somewhat mitigated by the skill of an intelligent medical man, who, even if he happens to know little about the habit of opium-eating, should know much as to the proper regimen to be observed in cases where these organs are disordered.

In respect to food it seems impossible to lay down any general rule. De Quincey advises beefsteak, not too much cooked, and stale bread as the chief diet, and doubtless this was the best diet for him. Yet it is not the less true that "what is one man's meat is another man's poison," and food that is absolutely harmless to one may disorder the entire digestion of another. Roast pork, mince pies, and cheese do not, I believe, rank high with the Faculty for ease of digestion, yet I have found them comparatively innoxious, while poultry, milk, oysters, fish, some kinds of vegetables, and even dry toast have caused me serious inconvenience. The appetite of the recovering opium-eater will probably be voracious and not at all discriminating during the earlier stages of his experiment, and will continue unimpaired even when the stomach begins to be fastidious as to what it will receive. Probably no safer rule can be given than to limit the quantity eaten as far as practicable, and to use only such food as in each particular case is found to be most easy of digestion.

Too much prominence can not be given to bodily exercise as intimately connected with the recovery of the patient. Without this it seems to me doubtful whether a person could withstand the extreme irritation of his nervous system. In his worst state he can not sit still ; he must be moving. The complication of springs in the famous Kilmansegge leg, is nothing compared with the necesity for motion which is developed in the limbs of the recovering opium-eater. Whatever his health, whatever his spirits, whatever the weather, walk he must.   Ten miles before breakfast will be found a moderate allowance for many months after the habit has been subdued.   A patient who could afford to give up three months of his time after the opium had been entirely discarded, to the perfect recovery of his health, could probably turn it to no better account than by stretching out on a pedestrian excursion of a thousand miles and back. This would be at the rate of nearly twenty-six miles a day, allowing Sunday as a day of rest.   This advice is seriously given for the consideration of those who can command the time for such a thorough process of restoration.   Nor should any weight be given to the objection that the body is in too enfeebled a state to make it safe to venture upon such an experiment.   Account for it as physiologists may, it is certain that the debilitating effects of leaving off opium much more rapidly pass away from the lower extremities than from the rest of the body.   At no time subsequent to my mastery of opium have I found any difficulty in accomplishing the longest walks ; on the contrary they have been taken with entire ease and pleasure.   Yet to this day, any considerable exercise of the other muscles is attended with extreme debility.   In the absence of facilities for walking, gymnastic exercise is not wholly without benefit, and if this exercise is followed by a cold bath, some portion of the insupportable languor will be removed.   Walking, however, is the great panacea, nor can it well be taken in excess.   So important is this element in the restorative process that it may well be doubted whether without its aid a confirmed opium-eater could be restored to health.

It is useless for any person to think that he can break off even the least inveterate of his habits without effort, or the

more obstinate ones without a struggle. Wine, spirits, tobacco, after years of habitual use, require a degree of resolution which is sometimes found to be beyond the resources
of the will. Much more does opium, whose hold upon the
system is vastly more tenacious than all these combined, call
for a resolute determination prepared to meet all the possible consequences that pertain to a complete and perfect
mastery of the habit. It should be remembered, however,
that the experience here recorded is that resulting from years
of large and uninterrupted use of opium. The entire system
had necessarily conformed itself to the artificial habit. For
years the proper action of the nervous, muscular, digestive,
and secretory system had been impeded and forced in an
unnatural direction. In time all the vital functions had conformed as far as possible to the necessity imposed upon
them. Scarce a function of the body that had not been
daily drilled into a highly artificial adaptation to the conditions imposed upon the system by the use of opium. Nature,
indeed, for a time rebels and resists the attempt to impose
unnatural habitudes upon her action ; but there is a limit to
her resistance, and she is then found to possess a marvellous
power of reconciling the processes of life with the disturbance and disorder of almost the entire human organization.
This power of adaptation, while it unquestionably lures on
to the continued indulgence of all kinds of bad habits, is, on
the other hand, the only hope and assurance the sufferer
from such causes can have of ultimate recovery from his
danger. If it requires years to establish bad habits in the
animal economy, why should we expect that they can be
wholly eradicated except by a reversal, in these respects, of
the entire current of the life, or without allowing a commen-
surate time for that perfect restoration of the disordered
functions which is expected ?

If this view of the case is not encouraging to the veteran consumer of opium, it certainly is not without its suggestive utility to that larger class whose use of opium has been comparatively limited both in time and quantity. Fortunately, much the greater number of opium-eaters take the drug in small quantities or have made use of it for only a limited period. In their case the process of recovery is relatively easy ; the functions of their physical organization still act for the most part in a normal way ; they have to retrace comparatively few steps and for comparatively a short time. Even to the inveterate consumer of the drug it has been made manifest that he may emancipate himself from his bondage if he will manfully accept the conditions upon which alone he can accomplish it. In the worst conceivable cases it is at least a choice between evils ; if he abandons opium, he may count upon much suffering of body, many sleepless nights, a disordered nervous system, and at times great prostration of strength. If he continues the habit, there remains, as long as life lasts, the irresolute will, the bodily languor, the ever-present sense of hopeless, helpless ruin. The opium-eater must take his choice between the two. On the one hand is hope, continually brightening in the future—on the other is the inconceivable wretchedness of one from whom hope has forever fled.

DE QUINCEY'S "CONFESSIONS OF AN ENGLISH OPIUM-EATER."

UNDER this title an article appeared in the "London Magazine" for December, 1821, which attracted very general attention from its literary merit and the novelty of its revelations. So considerable was the interest excited in these "Confessions" that the article was speedily republished in book form both in London and this country. The reading public outside of the medical profession were thus for the first time made generally acquainted with the tremendous potency of a drug whose fascinations have since become almost as well known to the inhabitants of England and America as to the people of India or China. The general properties of the drug had of course been familiar to intelligent men from the days of Vasco de Gama, but how easily the habit of using it could be acquired, and with what difficulty when acquired it could be left off, were subjects respecting which great obscurity rested on the minds even of medical men. Such parts only of these "Confessions" as have relation to De Quincey's habits as an opium-eater, have been selected for republication; such extracts from his other writings are added as embody his entire experience of opium so far as he has given it to the world.

I here present you, courteous reader, with the record of a remarkable period of my life. According to my application of it, I trust that it will prove not merely an interesting record, but in a considerable degree useful and instructive. In

that hope it is that I have drawn it up, and *that* must be my
apology for breaking through that delicate and honorable
reserve which for the most part restrains us from the public
exposure of our own errors and infirmities.

Guilt and misery shrink by a natural instinct from public
notice : they court privacy and solitude ; and, even in the
choice of a grave, will sometimes sequester themselves from
the general population of the church-yard, as if declining to
claim fellowship with the great family of man, and wishing—
in the affecting language of Mr. Wordsworth—

> " Humbly to express
> A penitential loneliness."

It is well, upon the whole, and for the interest of us all
that it should be so ; nor would I willingly, in my own person, manifest a disregard of such salutary feelings, nor in
act or word do any thing to weaken them.   But on the one
hand, as my self-accusation does not amount to a confession
of guilt, so on the other, it is possible that, if it did, the benefit resulting to others from the record of an experience purchased at so heavy a price might compensate, by a vast over-
balance, for any violence done to the feelings I have noticed,
and justify a breach of the general rule.   Infirmity and misery do not, of necessity, imply guilt.   They approach or recede from the shades of that dark alliance in proportion to
the probable motives and prospects of the offender, and the
palliations, known or secret, of the offense ; in proportion as
the temptations to it were potent from the first, and the resistance to it, in act or in effort, was earnest to the last.   For
my own part, without breach of truth or modesty, I may affirm that my life has been on the whole the life of a philosopher ; from my birth I was made an intellectual creature ;
and intellectual in the highest sense my pursuits and pleasures have been, even from my school-boy days.   If opium-

eating be a sensual pleasure, and if I am bound to confess that I have indulged in it to an excess not yet *recorded** of any other man, it is no less true that I have struggled against this fascinating enthrallment with a religious zeal, and have at length accomplished what I never yet heard attributed to any other man—have untwisted, almost to its final links, the accursed chain which fettered me. Such a self-conquest may reasonably be set off in counterbalance to any kind or degree of self-indulgence. Not to insist that, in my case, the self-conquest was unquestionable, the self-indulgence open to doubts of casuistry, according as that name shall be extended to acts aiming at the bare relief of pain, or shall be restricted to such as aim at the excitement of positive pleasure.

Guilt, therefore, I do not acknowledge; and, if I did, it is possible that I might still resolve on the present act of confession, in consideration of the service which I may thereby render to the whole class of opium-eaters. But who are they? Reader, I am sorry to say, a very numerous class indeed. Of this I became convinced some years ago, by computing at that time the number of those in one small class of English society (the class of men distinguished for talent, or of eminent station) who were known to me, directly or indirectly, as opium-eaters; such, for instance, as the eloquent and benevolent ——, the late Dean of ——; Lord ——; Mr. ——, the philosopher; a late under-secretary of state (who described to me the sensation which first drove him to the use of opium in the very same words of the Dean of ——, viz., "that he felt as though rats were gnawing and abrading the coats of his stomach"); Mr. ——; and many others, hard-

---

* "Not yet *recorded*," I say; for there is one celebrated man of the present day [Coleridge] who, if all be true which is reported of him, has greatly exceeded me in quantity.

ly less known, whom it would be tedious to mention. Now
if one class, comparatively so limited, could furnish so many
scores of cases (and that within the knowledge of one single
inquirer), it was a natural inference that the entire population of England would furnish a proportionable number·
The soundness of this inference, however, I doubted, until
some facts became known to me which satisfied me that it
was not incorrect.   I will mention two : 1. Three respectable London druggists, in widely remote quarters of London,
from whom I happened lately to be purchasing small quantities of opium, assured me that the number of *amateur* opium-
eaters (as I may term them) was at this time immense ; and
that the difficulty of distinguishing these persons, to whom
habit had rendered opium necessary, from such as were purchasing it with a view to suicide, occasioned them daily
trouble and disputes.   This evidence respected London only.
But, 2, (which will possibly surprise the reader more,) some
years ago, on passing through Manchester, I was informed
by several cotton manufacturers that their work-people were
rapidly getting into the practice of opium-eating ; so much
so that on a Saturday afternoon the counters of the druggists
were strewed with pills of one, two, or three grains, in preparation for the known demand of the evening.   The immediate occasion of this practice was the lowness of wages,
which at that time would not allow them to indulge in ale or
spirits, and wages rising, it may be thought that this practice
would cease ; but as I do not readily believe that any man,
having once tasted the divine luxuries of opium, will afterward descend to the gross and mortal enjoyments of alcohol,
I take it for granted

> " That those eat now who never ate before ;
> And those who always ate, now eat the more."

I have often been asked how I first came to be a regular

opium-eater, and have suffered very unjustly in the opinion of my acquaintance, from being reputed to have brought upon myself all the sufferings which I shall have to record, by a long course of indulgence in this practice purely for the sake of creating an artificial state of pleasurable excitement. This, however, is a misrepresentation of my case. True it is that for nearly ten years I did occasionally take opium for the sake of the exquisite pleasure it gave me ; but, so long as I took it with this view, I was effectually protected from all material bad consequences by the necessity of interposing long intervals between the several acts of indulgence, in order to renew the pleasurable sensations. It was not for the purpose of creating pleasure, but of mitigating pain in the severest degree, that I first began to use opium as an article of daily diet. In the twenty-eighth year of my age a most painful affection of the stomach, which I had first experienced about ten years before, attacked me in great strength. This affection had originally been caused by the extremities of hunger suffered in my boyish days. During the season of hope and redundant happiness which succeeded (that is, from eighteen to twenty-four) it had slumbered ; for the three following years it had revived at intervals ; and now, under unfavorable circumstances, from depression of spirits, it attacked me with a violence that yielded to no remedies but opium.

It is so long since I first took opium, that if it had been a trifling incident in my life I might have forgotten its date ; but cardinal events are not to be forgotten ; and, from circumstances connected with it, I remember that it must be referred to the autumn of 1804. During that season I was in London, having come thither for the first time since my entrance at college. And my introduction to opium arose in the following way : From an early age I had been accustomed to wash my head in cold water at least once a day. Being suddenly seized with toothache, I attributed it to some relaxation caused by an accidental intermission of that practice ; jumped out of bed, plunged my head into a basin of cold water, and with hair thus wetted went to sleep. The next morning, as I need hardly say, I awoke with excruciating rheumatic pains of the head and face, from which I had hardly any respite for about twenty days. On the twenty-first day I think it was, and on a Sunday, that I went out into the streets ; rather to run away, if possible, from my torments than with any distinct purpose. By accident I met a college acquaintance, who recommended opium. Opium ! dread agent of unimaginable pleasure and pain ! I had heard of it as I had heard of manna or of ambrosia, but no further. How unmeaning a sound it was at that time ! what solemn chords does it now strike upon my heart ! what heart-quaking vibrations of sad and happy remembrances ! It was a Sunday afternoon, wet and cheerless ; and a duller spectacle this earth of ours has not to show than a rainy Sunday in London. My road homeward lay through Oxford Street, and near the " Pantheon " I saw a druggist's shop. The druggist (unconscious minister of celestial pleasures !), as if in sympathy with the rainy Sunday, looked dull and stupid, just as any mortal druggist might be expected to look on a Sunday, and when I asked for the tincture of opium he gave it to me as any other man might do ; and furthermore, out of my shilling returned to me what seemed to be a real copper half-penny, taken out of a real wooden drawer. Nevertheless, in spite of such indications of humanity, he has ever since existed in my mind as a beatific vision of an immortal druggist sent down to earth on a special mission to myself.

Arrived at my lodgings, it may be supposed that I lost

not a moment in taking the quantity prescribed. I was necessarily ignorant of the whole art and mystery of opium-taking, and what I took, I took under every disadvantage. But I took it; and in an hour—O heavens! what a revulsion! what an upheaving from its lowest depths of the inner spirit! what an apocalypse of the world within me! That my pains had vanished was now a trifle in my eyes—this negative effect was swallowed up in the immensity of those positive effects which had opened before me in the abyss of divine enjoyment thus suddenly revealed. Here was a panacea, a φαρμακον νεπενθες, for all human woes; here was the secret of happiness, about which philosophers had disputed for so many ages, at once discovered. Happiness might now be bought for a penny and carried in the waistcoat pocket; portable ecstacies might be had corked up in a pint-bottle; and peace of mind could be sent down in gallons by the mail-coach. But if I talk in this way the reader will think I am laughing, and I can assure him that nobody will laugh long who deals much with opium. Its pleasures even are of a grave and solemn complexion, and in his happiest state the opium-eater can not present himself in the character of *L'Allegro;* even then he speaks and thinks as becomes *Il Penseroso.*

And first one word with respect to its bodily effects; for upon all that has been hitherto written on the subject of opium, whether by travellers in Turkey (who may plead their privilege of lying as an old immemorial right) or by professors of medicine, writing *ex cathedra,* I have but one emphatic criticism to pronounce—Lies! lies! lies! I do by no means deny that some truths have been delivered to the world in regard to opium: thus it has been repeatedly affirmed by the learned that opium is a dusky brown in color, and this, take notice, I grant; secondly, that it is rather dear, which also I grant—for in my time East India opium has been three guineas a pound, and Turkey eight ; and thirdly, that if you eat a good deal of it, most probably you must do what is particularly disagreeable to any man of regular habits, viz., die. These weighty propositions are, all and singular, true ; I can not gainsay them ; and truth ever was and will be commendable. But in these three theorems I believe we have exhausted the stock of knowledge as yet accumulated by man on the subject of opium. And therefore, worthy doctors, as there seems to be room for further discoveries, stand aside and allow me to come forward and lecture on this matter.

First, then, it is not so much affirmed as taken for granted by all who ever mention opium, formally or incidentally, that it does or can produce intoxication. Now, reader, assure yourself, *meo periculo*, that no quantity of opium ever did or could intoxicate. As to the tincture of opium, commonly called laudanum, *that* might certainly intoxicate if a man could bear to take enough of it ; but why ? because it contains so much proof spirit, and not because it contains so much opium. But crude opium, I affirm peremptorily, is incapable of producing any state of body at all resembling that which is produced by alcohol ; and not in *degree* only incapable, but even in *kind ;* it is not in the quantity of its effects merely, but in the quality, that it differs altogether. The pleasure given by wine is always mounting and tending to a crisis, after which it declines ; that from opium, when once generated, is stationary for eight or ten hours ; the first, to borrow a technical distinction from medicine, is a case of acute, the second of chronic, pleasure ; the one is a flame, the other a steady and equable glow. But the main distinction lies in this, that whereas wine disorders the mental faculties, opium, on the contrary (if taken in a proper manner), introduces among them the most exquisite order, legislation, and harmony. Wine robs a man of his self-possession; opium greatly invigorates it. Wine unsettles and clouds the judgment, and gives a preternatural brightness and a vivid exaltation to the contempts and the admirations, to the loves and the hatreds, of the drinker; opium, on the contrary, communicates serenity and equipoise to all the faculties, active or passive; and, with respect to the temper and moral feelings in general, it gives simply that sort of vital warmth which is approved by the judgment, and which would proba-bly always accompany a bodily constitution of primeval or antediluvian health. Thus, for instance, opium, like wine, gives an expansion to the heart and the benevolent affec-tions; but then with this remarkable difference, that in the sudden development of kind-heartedness which accompanies inebriation there is always more or less of a maudlin charac-ter which exposes it to the contempt of the by-stander. Men shake hands, swear eternal friendship, and shed tears— no mortal knows why—and the sensual creature is clearly uppermost. But the expansion of the beniger feelings, in-cident to opium, is no febrile access, but a healthy restoration to that state which the mind would naturally recover upon the removal of any deep-seated irritation of pain that had disturbed and quarrelled with the impulses of a heart origi-nally just and good. Wine constantly leads a man to the brink of absurdity and extravagance, and beyond a certain point it is sure to volatilize and to disperse the intellectual energies; whereas opium always seems to compose what had been agitated, and to concentrate what had been dis-tracted. In short, to sum up all in one word, a man who is inebriated, or tending to inebriation, is, and feels that he is, in a condition which calls up into supremacy the merely hu-màn, too often the brutal, part of his nature; but the opium-eater (I speak of him who is not suffering from any disease, or other remote effects of opium) feels that the diviner part of his nature is paramount ; that is, the moral affections are in a state of cloudless serenity ; and over all is the great light of the majestic intellect.

This is the doctrine of the true Church on the subject of opium : of which Church I acknowledge myself to be the only member—the alpha and omega ; but then it is to be recollected that I speak from the ground of a large and profound personal experience, whereas most of the unscientific authors who have at all treated of opium, and even of those who have written expressly on the *materia medica*, make it evident from the horror they express of it that their experimental knowledge of its action is none at all. I will, however, candidly acknowledge that I have met with one person who bore evidence to its intoxicating power such as staggered my own incredulity; for he was a surgeon, and had himself taken opium largely. I happened to say to him, that his enemies (as I had heard) charged him with talking nonsense on politics, and that his friends apologized for him by suggesting that he was constantly in a state of intoxication from opium. Now the accusation, said I, is not *prima facie*, and of necessity an absurd one ; but the defense *is*. To my surprise, however, he insisted that both his enemies and his friends were in the right. "I will maintain," said he, "that I *do* talk nonsense ; and secondly, I will maintain that I do not talk nonsense upon principle, or with any view to profit, but solely and simply," said he, "solely and simply—solely and simply," repeating it three times over, "because I am drunk with opium ; and that daily." I confess, that the authority of a surgeon, and one who was reputed a good one, may seem a weighty one to my prejudice ; but still I must plead my experience, which was greater than his greatest by

seven thousand drops a day ; and though it was not possible
to suppose a medical man unacquainted with the characteristic symptoms of vinous intoxication, yet it struck me that he
might proceed on a logical error of using the word intoxication with too great latitude, and extending it generically to
all modes of nervous excitement, instead of restricting it as
the expression for a specific sort of excitement connected
with certain diagnostics. Some people have maintained, in
my hearing, that they had been drunk upon green tea ; and
a medical student in London, for whose knowledge in his
profession I have reason to feel great respect, assured me
the other day that a patient in recovering from an illness
had got drunk on a beefsteak.

Having dwelt so much on this first and leading error in
respect to opium, I shall notice very briefly a second and a
third ; which are, that the elevation of spirits produced by
opium is necessarily followed by a proportionate depression,
and that the natural and even immediate consequence of
opium is torpor and stagnation, animal and mental. The
first of these errors I shall content myself with simply deny-
ing ; assuring my reader that for ten years, during which I
took opium at intervals, the day succeeding to that on which
I allowed myself this luxury was always a day of unusually
good spirits.

With respect to the torpor supposed to follow, or rather
(if we were to credit the numerous pictures of Turkish opium-eaters) to accompany the practice of opium-eating, I deny
that also. Certainly, opium is classed under the head of
narcotics, and some such effect it may produce in the end,
but the primary effects of opium are always, and in the highest degree, to excite and stimulate the system. This first
stage of its action always lasted with me, during my novitiate,
for upward of eight hours, so that it must be the fault of

the opium-eater himself if he does not so time his exhibition
of the dose (to speak medically) as that the whole weight of
its narcotic influence may descend upon his sleep.

Thus I have shown that opium does not, of necessity, produce inactivity or torpor. On the contrary it often led me
into markets and theatres. Yet, in candor, I will admit that
markets and theatres are not the appropriate haunts of the
opium-eater when in the divinest state incident to his enjoyment. In that state crowds become an oppression to him ;
music, even, too sensual and gross. He naturally seeks solitude and silence as indispensable conditions of those trances,
or profoundest reveries, which are the crown and consumma-
tion of what opium can do for human nature.

Courteous, and I hope indulgent reader, having accompanied me thus far, now let me request you to move onward
for about eight years ; that is to say, from 1804 (when I said
that my acquaintance with opium first began) to 1812.
And what am I doing? Taking opium. Yes, but what else?
Why, reader, in 1812, the year we are now arrived at, as well
as for some years previous, I have been chiefly studying
German metaphysics, in the writings of Kant, Fichte, Schel-
ling, etc. And I still take opium? On Saturday nights. And,
perhaps, have taken it unblushingly ever since " the rainy
Sunday," and " the Pantheon," and " the beatific druggist "
of 1804? Even so. And how do I find my health after all
this opium-eating? in short, how do I do? Why, pretty well,
I thank you, reader ; in the phrase of ladies in the straw, " as
well as can be expected." In fact, if I dared to say the real and
simple truth (it must not be forgotten that hitherto I thought,
to satisfy the theories of medical men, I ought to be ill), I
was never better in my life than in the spring of 1812 ; and
I hope sincerely that the quantity of claret, port, or " particular Maderia," which in all probability you, good reader,

have taken and design to take for every term of eight years during your natural life, may as little disorder your health as mine was disordered by opium I·had taken for the eight years between 1804 and 1812. To this moderation and temperate use of the article I may ascribe it, I suppose, that as yet at least (that is, in 1812) I am ignorant and unsuspicious of the avenging terrors which opium has in store for those who abuse its lenity. At the same time I have been only a *dilettante* eater of opium ; eight years' practice even, with the single precaution of allowing sufficient intervals between every indulgence, has not been sufficient to make opium necessary to me as an article of daily diet. But now comes a different era. Move on, if you please, reader, to 1813. In the summer of the year we have just quitted I had suffered much in bodily health from distress of mind connected with a very melancholy event. This event, being no ways related to the subject now before me further than through bodily illness which it produced, I need not more particularly notice. Whether this illness of 1812 had any share in that of 1813 I know not ; but so it was, that in the latter year I was attacked by a most appalling irritation of the stomach, in all respects the same as that which had caused me so much suffering in youth, and accompanied by a revival of all the old dreams. This is the point of my narrative on which, as respects my own self-justification, the whole of what follows may be said to hinge. And here I find myself in a perplexing dilemma. Either, on the one hand, I must exhaust the reader's patience by such a detail of my malady and of my struggles with it as might suffice to establish the fact of my inability to wrestle any longer with irritation and constant suffering, or, on the other hand, by passing lightly over this critical part of my story, I must forego the benefit of a stronger impression left on the mind of the reader, and

must lay myself open to the misconstruction of having slipped by the easy and gradual steps of self-indulging persons
from the first to the final state of opium-eating (a misconstruction to which there will be a lurking predisposition in
most readers from my previous acknowledgments).   Be not
so ungenerous as to let me suffer in your good opinion
through my own forbearance and regard for your comfort.
No ; believe all that I ask of you, viz., that I could resist no
longer.   Whether, indeed, afterward, I might not have succeeded in breaking off the habit, even when it seemed to me
that all efforts would be unavailing, and whether many of
the innumerable efforts which I *did* make might not have
been carried much further, and my gradual re-conquests of
ground lost might not have been followed up much more energetically, these are questions which I must decline.   Perhaps I might make out a case of palliation ; but—shall I
speak ingenuously ?—I confess it, as a besetting infirmity of
mine, that I am too much of an Eudæmonist ; I hanker too
much after a state of happiness, both for myself and others ;
I can not face misery, whether my own or not, with an eye
of sufficient firmness ; and am little capable of encountering
present pain for the sake of any reversionary benefit.

The issue of the struggle in 1813 was what I have mentioned ; and from this date the reader is to consider me as a
regular and confirmed opium-eater, of whom to ask whether
on any particular day he had or had not taken opium, would
be to ask whether his lungs had performed respiration, or the
heart fulfilled its functions.   Now then, reader, from 1813,
where all this time we have been sitting down and loitering,
rise up, if you please, and walk forward about three years
more.   Now draw up the curtain, and you shall see me in a
new character.

This year which we have now reached, stood, I confess,

as a parenthesis between years of a gloomier character. It was a year of brilliant water (to speak after the manner of jewellers), set, as it were, and insulated in the gloom and cloudy melancholy of opium. Strange as it may sound, I had a little before this time descended suddenly, and without any considerable effort, from three hundred and twenty grains of opium (that is, eight* thousand drops of laudanum) per day to forty grains, or one-eighth part. Instantaneously, and as if by magic, the cloud of profoundest melancholy which rested upon my brain, like some black vapors that I have seen roll away from the summits of mountains, drew off in one day ; passed off with its murky banners as simultaneously as a ship that has been stranded and is floated off by a spring tide—

"That moveth altogether, if it move at all."

Now, then, I was again happy. I now took only one thousand drops of laudanum per day—and what was that? A latter spring had come to close up the season of youth. My brain performed its functions as healthily as ever before. I read Kant again, and again I understood him, or fancied that I did. Again my feelings of pleasure expanded themselves to all around me. And, by the way, I remember about this time a little incident, which I mention because trifling as it was the reader will soon meet it again in my dreams, which it influenced more fearfully than could be imagined. One day a Malay knocked at my door. What business a Malay

* I here reckon twenty-five drops of laudanum as equivalent to one grain of opium, which I believe is the common estimate. However, as both may be considered variable quantities (the crude opium varying much in strength, and the tincture still more), I suppose that no infinitesimal accuracy can be had in such a calculation. Tea-spoons vary as much in size as opium in strength. Small ones hold about one hundred drops—so that eight thousand drops are about eighty times a tea-spoonful.

could have to transact among English mountains I can not conjecture, but possibly he was on his road to a sea-port about forty miles distant.

The servant who opened the door to him was a young girl born and bred among the mountains, who had never seen an Asiatic dress of any sort. His turban, therefore, confounded her not a little; and as it turned out that his attainments in English were exactly of the same extent as hers in the Malay, there seemed to be an impassable gulf fixed between all communication of ideas, if either party had happened to possess any. In this dilemma, the girl, recollecting the reputed learning of her master (and doubtless giving me credit for a knowledge of all the languages of the earth, besides perhaps a few of the lunar ones), came and gave me to understand that there was a sort of demon below whom she clearly imagined that my art could exorcise from the house. I did not immediately go down, but when I did the group which presented itself—arranged as it was by accident—though not very elaborate, took hold of my fancy and my eye in a way that none of the statuesque attitudes exhibited in the ballets at the opera-house, though so ostentatiously complex, had ever done. In a cottage kitchen, but panelled on the wall with dark wood that from age and rubbing resembled oak, and looking more like a rustic hall of entrance than a kitchen, stood the Malay, his turban and loose trowsers of dingy white relieved upon the dark panelling. He had placed himself nearer to the girl than she seemed to relish, though her native spirit of mountain intrepidity contended with the feeling of simple awe which her countenance expressed as she gazed upon the tiger-cat before her. And a more striking picture there could not be imagined than the beautiful English face of the girl, and its exquisite fairness, together with her erect and independent attitude, contrasted with the sallow and bilious skin of the Malay, enamelled or veneered with mahogany by marine air, his small, fierce, restless eyes, thin lips, slavish gestures, and adorations.   Half hidden by the ferocious-looking Malay was a little child from a neighboring cottage, who had crept in after him and was now in the act of reverting its head and gazing upward at the turban and the fiery eyes beneath it, while with one hand he caught at the dress of the young woman for protection.

My knowledge of the Oriental tongues is not remarkably extensive, being, indeed, confined to two words—the Arabic word for barley and the Turkish for opium (madjoon), which I have learned from Anastasius — and as I had neither a Malay dictionary, nor even Adelung's " Mithridates," which might have helped me to a few words, I addressed him in some lines from the Iliad ; considering that of such language as I possessed, the Greek, in point of longitude, came geographically nearest to an Oriental one.   He worshiped me in a devout manner, and replied in what I suppose was Malay.   In this way I saved my reputation with my neighbors, for the Malay had no means of betraying the secret. He lay down upon the floor for about an hour and then pursued his journey.   On his departure I presented him with a piece of opium.   To him, as an Orientalist, I concluded that opium must be familiar, and the expression of his face convinced me that it was.   Nevertheless, I was struck with some little consternation when I saw him suddenly raise his hand to his mouth, and (in the school-boy phrase) bolt the whole, divided into three pieces, at one mouthful.   The quantity was enough to kill three dragoons and their horses, and I felt some alarm for the poor creature.   But what could be done ? I had given him the opium in compassion for his solitary life, on recollecting that if he had travelled on foot from London it must be nearly three weeks since he could have exchanged a thought with any human being. I could not think of violating the laws of hospitality by having him seized and drenched with an emetic, and thus frightening him into a notion that we were going to sacrifice him to some English idol. No; there was clearly no help for it. He took his leave, and for some days I felt anxious; but as I never heard of any Malay being found dead, I became convinced that he was used * to opium, and that I must have done him the service I designed by giving him one night of respite from the pains of wandering.

This incident I have digressed to mention because this Malay (partly from the picturesque exhibition he assisted to frame, partly from the anxiety I connected with his image for some days) fastened afterward upon my dreams, and brought other Malays with him, worse than himself, that ran " a-muck "† at me, and led me into a world of troubles.

And now, reader, we have run through all the ten categories of my condition as it stood about 1816–1817, up to the middle of which latter year I judge myself to have been a happy man.

But now farewell, a long farewell to happiness, winter or summer! farewell to smiles and laughter! farewell to peace of mind! farewell to hope and to tranquil dreams, and to

---

* This, however, is not a necessary conclusion; the varieties of effect produced by opium on different constitutions are infinite. A London magistrate (Harriot's "Struggles through Life," vol. iii. p. 391, third edition) has recorded that, on the first occasion of his trying laudanum for the gout, he took FORTY drops, the next night SIXTY, and on the fifth night EIGHTY, without any effect whatever, and this at an advanced age. I have an anecdote from a country surgeon, however, which sinks Mr. Harriot's case into a trifle.

† See the common accounts, in any Eastern traveller or voyager, of the frantic excesses committed by Malays who have taken opium or are reduced to desperation by ill luck at gambling.

the blessed consolations of sleep! For more than three years and a half I am summoned away from these. I am now arrived at an Iliad of woes, for I have now to record *the pains of opium.*

Reader, who have thus far accompanied me, I must request your attention to a brief explanatory note on three points :

1. For several reasons I have not been able to compose the notes for this part of my narrative into any regular and connected shape. I give the notes disjointed as I find them, or have now drawn them up from memory. Some of them point to their own date, some I have dated, and some are undated. Whenever it could answer my purpose to transplant them from the natural or chronological order. I have not scrupled to do so. Sometimes I speak in the present, sometimes in the past tense. Few of the notes, perhaps, were written exactly at the period of time to which they relate ; but this can little affect their accuracy, as the impressions were such that they can never fade from my mind. Much has been omitted. I could not, without effort, constrain myself to the task of either recalling or constructing into a regular narrative the whole burden of horrors which lies upon my brain. This feeling partly I plead in excuse, and partly that I am now in London, and am a helpless sort of person who can not even arrange his own papers without assistance, and I am separated from the hands which are wont to perform for me the offices of an amanuensis.

2. You will think, perhaps, that I am too confidential and communicative of my own private history. It may be so. But my way of writing is rather to think aloud and follow my own humors than much to consider who is listening to me ; and if I stop to consider what is proper to be said to this or that person, I shall soon come to doubt whether any

E

part at all is proper. The fact is, I place myself at a distance of fifteen or twenty years ahead of this time, and suppose myself writing to those who will be interested about me hereafter ; and wishing to have some record of a time, the entire history of which no one can know but myself, I do it as fully as I am able with the efforts I am now capable of making because I know not whether I can ever find time to do it again.

3. It will occur to you often to ask, Why did I not release myself from the horrors of opium by leaving it off or diminishing it ? To this I must answer briefly—it might be supposed that I yielded to the fascinations of opium too easily ; it can not be supposed that any man can be charmed by its terrors. The reader may be sure, therefore, that I made attempts innumerable to reduce the quantity. I add, that those who witnessed the agonies of those attempts, and not myself, were the first to beg me to desist. But could not I have reduced it a drop a day, or by adding water have bisected or trisected a drop ? A thousand drops bisected would thus have taken nearly six years to reduce, and that would certainly not have answered. But this is a common mistake of those who know nothing of opium experimentally. I appeal to those who do, whether it is not always found that down to a certain point it can be reduced with ease and even pleasure, but that after that point further reduction causes intense suffering. Yes, say many thoughtless persons, who know not what they are talking of, you will suffer a little low spirits and dejection for a few days. I answer, no ; there is nothing like low spirits ; on the contrary, the mere animal spirits are uncommonly raised, the pulse is improved, the health is better. It is not there that the suffering lies. It has no resemblance to the sufferings caused by renouncing wine. It is a state of unutterable

irritation of stomach (which surely is not much like dejection), accompanied by intense perspirations, and feelings such as I shall not attempt to describe without more space at my command.

I shall now enter " *in medias res,*" and shall anticipate, from a time when my opium pains might be said to be at their *acme,* an account of their palsying effects on the intellectual faculties.

My studies have now been long interrupted. I can not read to myself with any pleasure, hardly with a moment's endurance ; yet I read aloud sometimes for the pleasure of others, because reading is an accomplishment of mine— and in the slang use of the word *accomplishment,* as a superficial and ornamental attainment, almost the only one I possess—and formerly, if I had any vanity at all connected with any endowment or attainment of mine, it was with this, for I had observed that no accomplishment was so rare. Of late, if I have felt moved by any thing in books, it has been by the grand lamentations of Sampson Agonistes, or the great harmonies of the Satanic speeches in " Paradise Regained," when read aloud by myself.

For nearly two years I believe that I read no book but one ; and I owe it to the author, in discharge of a great debt of gratitude, to mention what that was. The sublimer and more passionate poets I still read, as I have said, by snatches and occasionally, but my proper vocation, as I well knew, was the exercise of the analytic understanding. Now, for the most part, analytic studies are continuous, and not to be pursued by fits and starts, or fragmentary efforts. Mathematics, for instance, intellectual philosophy, etc., were all become insupportable to me ; I shrunk from them with a sense of powerless and infantine feebleness that gave me an anguish the greater from remembering the time when I grap-

pled with them to my own hourly delight; and for this further reason, because I had devoted the labor of my whole
life, and had dedicated my intellect, blossoms, and fruits to
the slow and elaborate toil of constructing one single work,
to which I had presumed to give the title of an unfinished
work of Spinoza's, viz., " *De Emendatione Humani Intellect-
us.*" This was now lying locked up, as by frost, like any
Spanish bridge or aqueduct, begun upon too great a scale
for the resources of the architect; and, instead of surviving
me as a monument of wishes at least, and aspirations, and a
life of labor dedicated to the exaltation of human nature in
that way in which God had best fitted me to promote so
great an object, it was likely to stand a memorial to my
children of hopes defeated, of baffled efforts, of materials
uselessly accumulated, of foundations laid that were never to
support a superstructure, of the grief and the ruin of the
architect. In this state of imbecility I had for amusement
turned my attention to political economy. In 1819 a friend
in Edinburgh sent me down Mr. Ricardo's book; and, recurring to my own prophetic anticipation of the advent of
some legislator for this science, I said, before I had finished
the first chapter, " Thou art the man !" Wonder and curios-
ity were emotions that had long been dead in me. Yet I
wondered once more : I wondered at myself that I could
once again be stimulated to the effort of reading ; and much
more I wondered at the book.

Thus did one simple work of profound understanding avail
to give me a pleasure and an activity which I had not
known for years—it roused me even to write, or at least to
dictate what M. wrote for me. It seemed to me that some
important truths had escaped even " the inevitable eye " of
Mr. Ricardo ; and as these were for the most part of such a
nature that I could express or illustrate them more briefly

and elegantly by algebraic symbols than in the usual clumsy and loitering diction of economists, the whole would not have filled a pocket-book; and being so brief, with M. for my amanuensis, even at this time, incapable as I was of all general exertion, I drew up my "Prolegomena to all Future Systems of Political Economy." I hope it will not be found redolent of opium; though, indeed, to most people, the subject itself is a sufficient opiate.

This exertion, however, was but a temporary flash, as the sequel showed; for I designed to publish my work. Arrangements were made at a provincial press about eighteen miles distant for printing it. An additional compositor was retained for some days on this account. The work was even twice advertised, and I was, in a manner, pledged to the fulfillment of my intention. But I had a preface to write, and a dedication—which I wished to make a splendid one—to Mr. Ricardo. I found myself quite unable to accomplish all this. The arrangements were countermanded, the compositor dismissed, and my "Prolegomena" rested peacefully by the side of its elder and more dignified brother.

I have thus described and illustrated my intellectual torpor in terms that apply, more or less, to every part of the four years during which I was under the Circean spells of opium. But for misery and suffering, I might, indeed, be said to have existed in a dormant state. I seldom could prevail on myself to write a letter; an answer of a few words to any that I received was the utmost that I could accomplish, and often *that* not until the letter had lain weeks, or even months, on my writing-table. Without the aid of M. all records of bills paid, or *to be* paid, must have perished, and my whole domestic economy—whatever became of Political Economy—must have gone into irretrievable confusion. I shall not afterward allude to this part of the case.

It is one, however, which the opium-eater will find in the end as oppressive and tormenting as any other, from the sense of incapacity and feebleness, from the direct embarrassments incident to the neglect or procrastination of each day's appropriate duties, and from the remorse which must often exasperate the stings of these evils to a reflective and conscientious mind. The opium-eater loses none of his moral sensibilities or aspirations ; he wishes and longs as earnestly as ever to realize what he believes possible, and feels to be exacted by duty ; but his intellectual apprehension of what is possible infinitely outruns his power, not of execution only, but even of power to attempt. He lies under the weight of incubus and nightmare ; he lies in sight of all that he would fain perform, just as a man forcibly confined to his bed by the mortal languor of a relaxing disease, who is compelled to witness injury or outrage offered to some object of his tenderest love : he curses the spells which chain him down from motion ; he would lay down his life if he might but get up and walk ; but he is powerless as an infant, and can not even attempt to rise.

I now pass to what is the main subject of these latter confessions, to the history and journal of what took place in my dreams ; for these were the immediate and proximate cause of my acutest suffering.

The first notice I had of any important change going on in this part of my physical economy was from the re-awaking of a state of eye generally incident to childhood or exalted states of irritability. I know not whether my reader is aware that many children, perhaps most, have a power of painting, as it were, upon the darkness, all sorts of phantoms. In some that power is simply a mechanic affection of the eye ; others have a voluntary or semi-voluntary power to dismiss or summon them ; or as a child once said to me when

I questioned him on this matter, " I can tell them to go, and they go ; but sometimes they come when I don't tell them to come." Whereupon I told him that he had almost as un-limited a command over apparitions as a Roman centurion over his soldiers. In the middle of 1817, I think it was, that this faculty became positively distressing to me. At night, when I lay awake in bed, vast processions passed along in mournful pomp ; friezes of never-ending stories, that to my feelings were as sad and solemn as if they were stories drawn from times before Œdipus or Priam, before Tyre, be-fore Memphis. And at the same time a corresponding change took place in my dreams ; a theatre seemed sudden-ly opened and lighted up within my brain, which presented nightly spectacles of more than earthly splendor. And the four following facts may be mentioned as noticeable at this time :

I. That as the creative state of the eye increased, a sym-pathy seemed to arise between the waking and the dreaming states of the brain in one point—that whatsoever I happened to call up and to trace by a voluntary act upon the darkness was very apt to transfer itself to my dreams, so that I fear-ed to exercise this faculty.

II. For this, and all other changes in my dreams, were accompanied by deep-seated anxiety and gloomy melancholy, such as are wholly incommunicable by words. I seemed every night to descend, not metaphorically, but literally to descend, into chasms and sunless abysses, depths below depths, from which it seemed hopeless that I could ever re-ascend. Nor did I, by waking, feel that I had re-ascended. This I do not dwell upon, because the state of gloom which attended these gorgeous spectacles—amounting at last to utter darkness, as of some suicidal despondency—can not be approached by words.

III. The sense of space, and in the end the sense of time, were both powerfully affected. Buildings, landscapes, etc., were exhibited in proportions so vast as the bodily eye is not fitted to receive. Space swelled and was amplified to an extent of unutterable infinity. This, however, did not disturb me so much as the vast expansion of time. I sometimes seemed to have lived for seventy or one hundred years in one night; nay, sometimes had feelings representative of a millennium passed in that time, or, however, of a duration far beyond the limits of any human experience.

IV. The minutest incidents of childhood, or forgotten scenes of later years, were often revived. I could not be said to recollect them, for if I had been told of them when waking I should not have been able to acknowledge them as parts of my past experience; but placed as they were before me, in dreams like intuitions, and clothed in all their evanescent circumstances and accompanying feelings, I *recognized* them instantaneously. I was once told by a near relative of mine, that having in her chilhdood fallen into a river, and being on the very verge of death but for the critical assistance which reached her, she saw in a moment her whole life, in its minutest incidents, arrayed before her simultaneously as in a mirror; and she had a faculty developed as suddenly for comprehending the whole and every part. This, from some opium experiences of mine, I can believe. I have, indeed, seen the same thing asserted twice in modern books, and accompanied by a remark which I am convinced is true, viz., that the dread book of account which the Scriptures speak of is in fact the mind itself of each individual. Of this, at least, I feel assured, that there is no such thing as *forgetting* possible to the mind. A thousand accidents may and will interpose a veil between our present consciousness and the secret inscriptions on the mind; accidents of the same sort will also rend away this veil; but alike, whether veiled or unveiled, the inscription remains forever — just as the stars seem to withdraw before the common light of day, whereas in fact we all know that it is the light which is drawn over them as a veil, and that they are waiting to be revealed when the obscuring day-light shall have withdrawn.

And now came a tremendous change, which, unfolding itself slowly like a scroll through many months, promised an abiding torment; and, in fact, it never left me until the winding up of my case. Hitherto the human face had often mixed in my dreams—but not despotically, nor with any special power of tormenting—but now that which I have called the tyranny of the human face began to unfold itself. Perhaps some part of my London life might be answerable for this. Be that as it may, now it was that upon the rocking waters of the ocean the human face began to appear; the sea appeared paved with innumerable faces, upturned to the heavens; faces, imploring, wrathful, despairing, surged upward by thousands, by myriads, by generations, by centuries: my agitation was infinite, my mind tossed, and surged with the ocean.

May, 1818.—The Malay has been a fearful enemy for months. I have been every night, through his means, transported into Asiatic scenes. Under the connecting feeling of tropical heat and vertical sunlights I brought together all creatures, birds, beasts, reptiles, all trees and plants, usages and appearances, that are found in all tropical regions, and assembled them together in China or Indostan. From kindred feelings I soon brought Egypt and all her gods under the same law. I was stared at, hooted at, grinned at, chattered at by monkeys, by paroquets, by cockatoos. I ran into pagodas, and was fixed for centuries at the summit or in secret rooms : I was the idol ; I was the priest ; I was worshiped ; I was sacrificed. I fled from the wrath of Bramah through all the forests of Asia : Vishnu hated me ; Seeva laid wait for me. I came suddenly upon Isis and Osiris : I had done a deed, they said, which the ibis and the crocodile trembled at. I was buried for a thousand years in stone coffins, with mummies and sphinxes, in narrow chambers at the heart of eternal pyramids. I was kissed with cancerous kisses by crocodiles, and laid, confounded with all unutterable slimy things, among reeds and Nilotic mud.

I thus give the reader some slight abstraction of my Oriental dreams, which always filled me with such amazement at the monstrous scenery that horror seemed absorbed for a while in sheer astonishment. Sooner or later came a reflux of feeling that swallowed up the astonishment and left me not so much in terror as in hatred and abomination of what I saw. Over every form, and threat, and punishment, and dim sightless incarceration, brooded a sense of eternity and infinity that drove me into an oppression as of madness. Into these dreams only it was, with one or two slight exceptions, that any circumstances of physical horror entered. All before had been moral and spiritual terrors. But here the main agents were ugly birds, or snakes, or crocodiles, especially the last. The cursed crocodile became to me the object of more horror than almost all the rest. I was compelled to live with him, and (as was always the case almost in my dreams) for centuries. I escaped sometimes, and found myself in Chinese houses with cane tables, etc. All the feet of the tables, sofas, etc., soon became instinct with life. The abominable head of the crocodile and his leering eyes looked out at me multiplied into a thousand repetitions, and I stood loathing and fascinated. And so often did this hideous reptile haunt my dreams that many times the very

same dream was broken up in the very same way : I heard
gentle voices speaking to me (I hear every thing when I am
sleeping), and instantly I awoke. It was broad noon, and
my children were standing hand in hand at my bedside,
come to show me their colored shoes, or new frocks, or to
let me see them dressed for going out. I protest that so
awful was the transition from the damned crocodile and the
other unutterable monsters and abortions of my dreams to
the sight of innocent *human* natures and of infancy, that in
the mighty and sudden revulsion of mind I wept, and could
not forbear it, as I kissed their faces.

It now remains that I should say something of the way
in which this conflict of horrors was finally brought to its
crisis. The reader is already aware that the opium-eater
has, in some way or other, " unwound, almost to its final
links, the accursed chain which bound him." By what
means ? To have narrated this according to the original intention would have far exceeded the space which can now
be allowed. It is fortunate, as such a cogent reason exists
for abridging it, that I should on a maturer view of the case
have been exceedingly unwilling to injure by any such unaf-
fecting details the impression of the history itself as an appeal to the prudence and the conscience of the yet uncon-
firmed opium-eater, or even (though a very inferior consideration) to injure its effect as a composition. The interest
of the judicious reader will not attach itself chiefly to the
subject of the fascinating spells, but to the fascinating power. Not the opium-eater, but the opium is the true hero of
the tale, and the legitimate centre on which the interest re-
volves. The object was to display the marvellous agency
of opium, whether for pleasure or for pain. If that is done,
the action of the piece has closed.

However, as some people in spite of all laws to the contrary will persist in asking what became of the opium-eater, and in what state he now is, I answer for him thus : The reader is aware that opium had long ceased to found its empire on spells of pleasure ; it was solely by the tortures connected with the attempt to abjure it that it kept its hold. Yet as other tortures, no less it may be thought, attended the non-abjuration of such a tyrant, a choice only of evils was left ; and *that* might as well have been adopted, which, however terrific in itself, held out a prospect of final restoration to happiness. This appears true ; but good logic gave the author no strength to act upon it. However, a crisis arrived for the author's life, and a crisis for other objects still dearer to him, and which will always be far dearer to him than his life, even now that it is again a happy one. I saw that I must die if I continued the opium. I determined, therefore, if that should be required, to die in throwing it off. How much I was at that time taking I can not say ; for the opium which I used had been purchased for me by a friend who afterward refused to let me pay him, so that I could not ascertain even what quantity I had used within a year. I apprehend, however, that I took it very irregularly, and that I varied from about fifty or sixty grains to one hundred and fifty a day. My first task was to reduce it to forty, to thirty, and, as fast as I could, to twelve grains.

I triumphed. But think not, reader, that therefore my sufferings were ended, nor think of me as of one sitting in a *dejected* state. Think of me as of one, even when four months had passed, still agitated, writhing, throbbing, palpitating, shattered ; and much, perhaps, in the situation of him who has been racked, as I collect the torments of that state from the affecting account of them left by a most innocent sufferer [William Lithgow] of the time of James I. Meantime I derived no benefit from any medicine except one prescribed

to me by an Edinburgh surgeon of great eminence, viz., am-
moniated tincture of valerian.   Medical account, therefore, of
my emancipation I have not much to give, and even that little,
as managed by a man so ignorant of medicine as myself,
would probably tend only to mislead.   At all events it would
be misplaced in this situation.   The moral of the narrative
is addressed to the opium-eater, and therefore of necessity
limited in its application.   If he is taught to fear and trem-
ble, enough has been effected.   But he may say that the
issue of my case is at least a proof that opium, after a seventeen years' use and an eight years' abuse of its powers, may
still be renounced ; and that he may chance to bring to the
task greater energy than I did, or that with a stronger constitution than mine he may obtain the same results with less.
This may be true.   I would not presume to measure the efforts of other men by my own.   I heartily wish him more
energy ; I wish him the same success.   Nevertheless, I had
motives external to myself which he may unfortunately want,
and these supplied me with conscientious supports which
mere personal interests might fail to supply to a mind debilitated by opium.

Jeremy Taylor conjectures that it may be as painful to be
born as to die.   I think it probable ; and during the whole
period of diminishing the opium I had the torments of a man
passing out of one mode of existence into another.   The
issue was not death, but a sort of physical regeneration, and
I may add that ever since, at intervals, I have had a restoration of more than youthful spirits, though under the pressure
of difficulties, which in a less happy state of mind I should
have called misfortunes.

One memorial of my former condition still remains : my
dreams are not yet perfectly calm ; the dread swell and agitation of the storm have not wholly subsided ; the legions

that encamped in them are drawing off, but not all departed ; my sleep is tumultuous, and like the gates of Paradise to our first parents when looking back from afar, it is still, in the tremendous line of Milton—

" With dreadful faces throng'd and fiery arms."

The preceding narrative was written by De Quincey in the summer of 1821. In December of the next year a further record of his experience was published in the form of the following *Appendix*.

Those who have read the " Confessions " will have closed them with the impression that I had wholly renounced the use of opium. This impression I meant to convey, and that for two reasons : first, because the very act of deliberately recording such a state of suffering necessarily presumes in the recorder a power of surveying his own case as a cool spectator, and a degree of spirits for adequately describing it which it would be inconsistent to suppose in any person speaking from the station of an actual sufferer ; secondly, because I, who had descended from so large a quantity as eight thousand drops to so small a one, comparatively speaking, as a quantity ranging between three hundred and one hundred and sixty drops, might well suppose that the victory was in effect achieved. In suffering my readers, therefore, to think of me as of a reformed opium-eater, I left no impression but what I shared myself, and, as may be seen, even this impression was left to be collected from the general tone of the conclusion and not from any specific words, which are in no instance at variance with the literal truth. In no long time after that paper was written I became sensible that the effort which remained would cost me far more energy than I had anticipated, and the necessity for making it was more appar-

ent every month.   In particular I became aware of an increasing callousness or defect of sensibility in the stomach,
and this I imagined might imply a scirrhous state of that organ either formed or forming.   An eminent physician, to
whose kindness I was at that time deeply indebted, informed
me that such a termination of my case was not impossible,
though likely to be forestalled by a different termination in
the event of my continuing the use of opium.   Opium, therefore, I resolved wholly to abjure as soon as I should find
myself at liberty to bend my undivided attention and energy
to this purpose.   It was not, however, until the 24th of June
last that any tolerable concurrence of facilities for such an
attempt arrived.   On that day I began my experiment, having previously settled in my own mind that I would not
flinch, but would "stand up to the scratch" under any possible "punishment."   I must premise that about one hundred and seventy or one hundred and eighty drops had been
my ordinary allowance for many months.   Occasionally I
had run up as high as five hundred, and once nearly to seven
hundred.   In repeated preludes to my final experiment I
had also gone as low as one hundred drops, but had found
it impossible to stand it beyond the fourth day, which, by the
way, I have always found more difficult to get over than any
of the preceding three.   I went off under easy sail—one
hundred and thirty drops a day for three days ; on the fourth
I plunged at once to eighty.   The misery which I now suffered "took the conceit" out of me at once, and for about a
month I continued off and on about this mark ; then I sunk
to sixty, and the next day to—none at all.   This was the
first day for nearly ten years that I had existed without
opium.   I persevered in my abstinence for ninety hours ;
that is, upward of half a week.   Then I took—ask me not
how much ; say, ye severest, what would ye have done ?

Then I abstained again ; then took about twenty-five drops ;
then abstained ; and so on.

Meantime the symptoms which attended my case for the
first six weeks of the experiment were these   enormous irritability and excitement of the whole system—the stomach, in
particular, restored to a full feeling of vitality and sensibility,
but often in great pain ; unceasing restlessness night and
day ; sleep—I scarcely knew what it was—three hours out
of the twenty-four was the utmost I had, and that so agitated and shallow that I heard every sound that was near
me ; lower jaw constantly swelling ; mouth ulcerated ; and
many other distressing symptoms that would be tedious to
repeat, among which, however, I must mention one because
it had never failed to accompany any attempt to renounce
opium, viz., violent sternutation.   This now became exceedingly troublesome ; sometimes lasting for two hours at once,
and recurring at least twice or three times a day.   I was not
much surprised at this, on recollecting what I had somewhere
heard or read, that the membrane which lines the nostrils is
a prolongation of that which lines the stomach, whence I believe are explained the inflammatory appearances about the
nostrils of dram-drinkers.   The sudden restoration of its
original sensibility to the stomach expressed itself, I suppose,
in this way.   It is remarkable, also, that during the whole
period of years through which I had taken opium I had never once caught cold—as the phrase is—nor even the slightest
cough.   But now a violent cold attacked me, and a cough
soon after.   In an unfinished fragment of a letter begun
about this time to ——, I find these words : "You ask me
to write the —— ——.   Do you know Beaumont and Fletch-
er's play of ' Thierry and Theodoret?'   There you will see my
case as to sleep ; nor is it much of an exaggeration in other
features.   I protest to you that I have a greater influx of

thoughts in one hour at present than in a whole year under the reign of opium. It seems as though all the thoughts which had been frozen up for a decade of years by opium, had now, according to the old fable, been thawed at once, such a multitude stream in upon me from all quarters. Yet such is my impatience and hideous irritability, that for one which I detain and write down fifty escape me. In spite of my weariness from suffering and want of sleep I can not stand still or sit for two minutes together. '*I nunc, et versus tecum meditare canoros.*' "

At this stage of my experiment I sent to a neighboring surgeon, requesting that he would come over to see me. In the evening he came, and after briefly stating the case to him I asked this question : Whether he did not think that the opium might have acted as a stimulus to the digestive organs, and that the present state of suffering in the stomach—which manifestly was the cause of the inability to sleep —might arise from indigestion ? His answer was, No : on the contrary, he thought that the suffering was caused by digestion itself, which should naturally go on below the consciousness, but which, from the unnatural state of the stomach, vitiated by so long a use of opium, was become distinctly perceptible. This opinion was plausible, and the unintermitting nature of the suffering disposes me to think that it was true ; for if it had been any mere *irregular* affection of the stomach it should naturally have intermitted occasionally, and constantly fluctuated as to degree. The intention of Nature, as manifested in the healthy state, obviously is to withdraw from our notice all the vital motions—such as the circulation of the blood, the expansion and contraction of the lungs, the peristaltic action of the stomach, etc.—and opium, it seems, is able in this as in other instances to counteract her purposes. By the advice of the surgeon I tried *bitters.*

For a short time these greatly mitigated the feelings under which I labored, but about the forty-second day of the experiment the symptoms already noticed began to retire and new ones to arise of a different and far more tormenting class. Under these, but with a few intervals of remission, I have since continued to suffer; but I dismiss them undescribed for two reasons : first, because the mind revolts from retracing circumstantially any sufferings from which it is removed by too short or by no interval. To do this with minuteness enough to make the review of any use would be indeed " *infandum renovare dolorem,*" and possibly without a sufficient motive ; for, secondly, I doubt whether this latter state be any way referable to opium, positively considered, or even negatively ; that is, whether it is to be numbered among the last evils from the direct action of opium or even among the earliest evils consequent upon a *want* of opium in a system long deranged by its use. Certainly one part of the symptoms might be accounted for from the time of year (August) ; for, though the summer was not a hot one, yet in any case the sum of all the heat *funded* (if one may say so) during the previous months, added to the existing heat of that month, naturally renders August in its better half the hottest part of the year ; and it so happened that the excessive perspiration which even at Christmas attends any great reduction in the daily quantum of opium, and which in July was so violent as to oblige me to use a bath five or six times a day, had about the setting in of the hottest season wholly retired, on which account any bad effect of the heat might be the more unmitigated. Another symptom, viz., what in my ignorance I call internal rheumatism (sometimes affecting the shoulders, etc., but more often appearing to be seated in the stomach), seemed again less probably attributable to the opium or the want of opium than to the dampness of the house which I inhabit, which had about that time attained its maximum, July having been as usual a month of incessant rain in our most rainy part of England.

Under these reasons for doubting whether opium had any connection with the latter stage of my bodily wretchedness— except indeed as an occasional cause, as having left the body weaker and more crazy, and thus predisposed to any mal-influence whatever—I willingly spare my reader all description of it. Let it perish to him ; and would that I could as easily say, let it perish to my own remembrances, that any future hours of tranquillity may not be disturbed by too vivid an ideal of possible human misery !

So much for the sequel of my experiment. As to the former stage, in which properly lies the experiment and its application to other cases, I must request my reader not to forget the reason for which I have recorded it. This was a belief that I might add some trifle to the history of opium as a medical agent. In this I am aware that I have not at all fulfilled my own intentions, in consequence of the torpor of mind, pain of body, and extreme disgust to the subject which besieged me while writing that part of my paper ; which part being immediately sent off to the press (distant about five degrees of latitude), can not be corrected or improved. But from this account, rambling as it may be, it is evident that thus much of benefit may arise to the persons most interest-ed in such a history of opium—viz., to opium-eaters in gen-eral—that it establishes for their consolation and encourage-ment the fact that opium may be renounced without greater sufferings than an ordinary resolution may support, and by a pretty rapid course of descent.

On which last notice I would remark that mine was *too* rapid, and the suffering therefore needlessly aggravated ; or rather perhaps it was not sufficiently continuous and equably graduated. But that the reader may judge for himself, and above all that the opium-eater who is preparing to retire from business may have every sort of information before him, I subjoin my diary.

FIRST WEEK.		Drops of Laud.	SECOND WEEK.		Drops of Laud.
Monday,	June 24	130	Monday,	July 1	80
Tuesday,	" 25	140	Tuesday,	" 2	80
Wednesday,	" 26	130	Wednesday,	" 3	90
Thursday,	" 27	80	Thursday,	" 4	100
Friday,	" 28	80	Friday,	" 5	80
Saturday,	" 29	80	Saturday,	" 6	80
Sunday,	" 30	80	Sunday,	" 7	80

THIRD WEEK.		Drops of Laud.	FOURTH WEEK.		Drops of Laud.
Monday,	July 8	300	Monday,	July 15	76
Tuesday,	" 9	50	Tuesday,	" 16	73½
Wednesday,	" 10 ⎫		Wednesday,	" 17	73½
Thursday,	" 11 ⎬ Hiatus in		Thursday,	" 18	70
Friday,	" 12 ⎪ MS.		Friday,	" 19	240
Saturday,	" 13 ⎭		Saturday,	" 20	80
Sunday,	" 14	76	Sunday,	" 21	350

FIFTH WEEK.		Drops of Laud.
Monday,	July 22	60
Tuesday,	" 23	none.
Wednesday,	" 24	none.
Thursday,	" 25	none.
Saturday,	" 27	none.
Friday,	" 26	200

What mean these abrupt relapses, the reader will ask, perhaps, to such numbers as 300, 350, etc.? The *impulse* to these relapses was mere infirmity of purpose; the *motive*, where any motive blended with the impulse, was either the principle of "*reculer pour mieux sauter*" (for under the torpor of a large dose, which lasted for a day or two, a less quantity satisfied the stomach, which on awaking found itself partly accustomed to this new ration), or else it was this principle— that of sufferings otherwise equal, those will be borne best which meet with a mood of anger. Now whenever I ascended to any large dose I was furiously incensed on the following day, and could then have borne any thing.

The narrative part of De Quincey's " Confessions " by no means exhausts the story of his suffering as recorded by himself. Scattered through his miscellaneous papers are to be found frequent references to the opium habit and its protracted hold upon the system long after the drug itself had been discarded. The succeeding extracts from his " Literary Reminiscences " will throw light upon his bodily and mental condition in the years immediately following his opium struggle :

" I was ill at that time and for years after—ill from the effects of opium upon the liver, and one primary indication of any illness felt in that organ is peculiar depression of spirits. Hence arose a singular effect of reciprocal action in maintaining a state of dejection. From the original physical depression caused by the derangement of the liver arose a sympathetic depression of the mind, disposing me to believe that I never *could* extricate myself; and from this belief arose, by reaction, a thousand-fold increase of the physical depression. I began to view my unhappy London life—a life of literary toils odious to my heart—as a permanent state of exile from my Westmoreland home. My three eldest children, at that time in the most interesting stages of childhood and infancy, were in Westmoreland, and so powerful was my feeling (derived merely from a deranged liver) of some long, never-ending separation from my family, that at length, in pure weakness of mind, I was obliged to relinquish my daily walks in Hyde Park and Kensington Gardens from

the misery of seeing children in multitudes that too forcibly
recalled my own.

" Meantime it is very true that the labors I had to face
would not even to myself, in a state of good bodily health,
have appeared alarming. *Myself,* I say, for in any state of
health I do not write with rapidity. Under the influence,
however, of opium, when it reaches its maximum in diseas-
ing the liver and deranging the digestive functions, all exertion whatever is revolting in excess. Intellectual exertion
above all is connected habitually, when performed under opium influence, with a sense of disgust the most profound for
the subject (no matter what) which detains the thoughts ; all
that morning freshness of animal spirits, which under ordinary circumstances consumes, as it were, and swallows up the
interval between one's self and one's distant object, all that
dewy freshness is exhaled and burned off by the parching effects of opium on the animal economy.

" I was, besides, and had been for some time engaged in
the task of unthreading the labyrinth by which I had reached, unawares, my present state of slavery to opium. I was
descending the mighty ladder, stretching to the clouds as it
seemed, by which I had imperceptibly attained my giddy al-
titude—that point from which it had seemed equally impossible to go forward or backward. To wean myself from opium I had resolved inexorably, and finally I accomplished
my vow. But the transition state was the worst state of all
to support. All the pains of martyrdom were there ; all the
ravages in the economy of the great central organ, the stomach, which had been wrought by opium ; the sickening disgust which attended each separate respiration ; and the
rooted depravation of the appetite and the digestion — all
these must be weathered for months upon months, and without stimulus (however false and treacherous) which, for some

part of each day, the old doses of laudanum would have supplied.   These doses were to be continually diminished, and
under this difficult dilemma : If, as some people advised, the
diminution were made by so trifling a quantity as to be imperceptible, in that case the duration of the process was in-
terminable and hopeless—thirty years would not have sufficed to carry it through.   On the other hand, if twenty-five
to fifty drops were withdrawn on each day (that is, from one
to two grains of opium), inevitably within three, four, or five
days the deduction began to tell grievously, and the effect
was to restore the craving for opium more keenly than ever.
There was the collision of both evils—that from the laudanum and that from the want of laudanum.   The last was a
state of distress perpetually increasing, the other was one
which did not sensibly diminish—no, not for a long period
of months.   Irregular motions, impressed by a potent agent
upon the blood and other processes of life, are slow to subside ; they maintain themselves long after the exciting cause
has been partially or even wholly withdrawn ; and, in my
case, they did not perfectly subside into the motion of
tranquil health for several years.   From all this it will be
easy to understand the *fact*—though after all impossible, without a similar experience, to understand the *amount*—of my
suffering and despondency in the daily task upon which circumstances had thrown me at this period—the task of writing and producing something for the journals, *invita Minerva*.
Over and above the principal operation of my suffering state,
as felt in the enormous difficulty with which it loaded every
act of exertion, there was another secondary effect which always followed as a reaction from the first.   And that this
was no accident or peculiarity attached to my individual temperament, I may presume from the circumstance that Mr.
Coleridge experienced the very same sensations, in the same

situation, throughout his literary life, and has often noticed
it to me with surprise and vexation. The sensation was
that of powerful disgust with any subject upon which he had
occupied his thoughts or had exerted his powers of composition for any length of time, and an equal disgust with the
result of his exertions—powerful abhorrence, I may call it,
absolute loathing òf all that he had produced.

" In after years Coleridge assured me that he never could
read any thing he had written without a sense of overpowering disgust.   Reverting to my own case, which was pretty
nearly the same as his, there was, however, this difference—
that at times, when I had slept at more regular hours for
several nights consecutively, and had armed myself by a sudden increase of the opium for a few days running, I recovered at times a remarkable glow of jovial spirits.   In some
such artificial respites, it was, from my usual state of distress, and purchased at a heavy price of subsequent suffering,
that I wrote the greater part of the opium ' Confessions ' in
the autumn of 1821.

" These circumstances I mention to account for my having written any thing in a happy or genial state of mind,
when I was in a general state so opposite, by my own description, to every thing like enjoyment.   That description,
as a *general* one, states most truly the unhappy condition,
and the somewhat extraordinary condition of feeling to
which opium had brought me.   I, like Mr. Coleridge, could
not endure what I had written for some time after I had
written it.   I also shrunk from treating any subject which I
had much considered ; but more, I believe, as recoiling from
the intricacy and the elaborateness which had been made
known to me in the course of considering it, and on account
of the difficulty or the toilsomeness which might be fairly
presumed from the mere fact that I *had* long considered it,

or could have found it necessary to do so, than from any blind mechanical feeling inevitably associated (as in Coleridge it was) with a second survey of the same subject. One other effect there was from the opium, and I believe it had some place in Coleridge's list of morbid affections caused by opium, and of disturbances extended even to the intellect, which was, that the judgment was for a time grievously impaired, sometimes even totally abolished, as applied to any thing I had recently written. Fresh from the labor of composition, I believe, indeed, that almost every man, unless he has had a very long and close experience in the practice of writing, finds himself a little dazzled and bewildered in computing the effect, as it will appear to neutral eyes, of what he has produced. But the incapacitation which I speak of here as due to opium, is of another kind and another degree. It is mere childish helplessness, or senile paralysis, of the judgment, which distresses the man in attempting to grasp the upshot and the total effect (the *tout ensemble*) of what he has himself so recently produced. There is the same imbecility in attempting to hold things steadily together, and to bring them under a comprehensive or unifying act of the judging faculty, as there is in the efforts of a drunken man to follow a chain of reasoning. Opium is said to have some *specific* effect of debilitation upon the memory :* that is, not merely the general one which might be supposed to accompany its morbid effects upon the bodily system, but

---

* The technical memory, or that which depends upon purely arbitrary links of connection, and therefore more upon a *nisus* or separate activity of the mind—that memory, for instance, which recalls names—is undoubtedly affected, and most powerfully, by opium. On the other hand, the *logical* memory, or that which recalls facts that are connected by fixed relations, and where A being given, B must go before or after—historical memory, for instance—is not much if at all affected by opium.

F

some other, more direct, subtle, and exclusive ; and this, of whatever nature, may possibly extend to the faculty of judging. Such, however, over and above the more known and more obvious ill effects upon the spirits and the health, were some of the stronger and more subtle effects of opium in disturbing the intellectual system as well as the animal, the functions of the will also no less than those of the intellect, from which both Coleridge and myself were suffering at the period to which I now refer (1821–25) ; evils which found their fullest exemplification in the very act upon which circumstances had now thrown me as the *sine qua non* of my extrication from difficulties—viz., the act of literary composition. This necessity—the fact of its being my one sole resource for the present, and the established experience which I now had of the peculiar embarrassments and counteracting forces which I should find in opium, but still more in the train of consequences left behind by past opium—strongly co-operated with the mere physical despondency arising out of the liver : and the state of partial unhappiness, among other outward indications, expressed itself by one mark, which some people are apt greatly to misapprehend—as if it were some result of a sentimental turn of feeling—I mean perpetual sighs. But medical men must very well know that a certain state of the liver, *mechanically* and without any co-operation of the will, expresses itself in sighs. I was much too firm-minded and too reasonable to murmur or complain. I certainly suffered deeply, as one who finds himself a banished man from all that he loves, and who had not the consolations of hope, but feared too profoundly that all my efforts—efforts poisoned so sadly by opium—might be unavailing for the end.

"In 1824 I had come up to London upon an errand—in itself sufficiently vexatious—of fighting against pecuniary em-

barrassments by literary labors; but, as always happened
hitherto, with very imperfect success, from the miserable
thwartings I incurred through the deranged state of the liver.
My zeal was great and my application was unintermitting,
but spirits radically vitiated, chiefly through the direct mechanical depression caused by one important organ deranged; and secondly, by a reflex effect of depression through
my own thoughts in estimating my prospects, together with
the aggravation of my case by the inevitable exile from my
own mountain home—all this reduced the value of my exertion in a deplorable way. It was rare, indeed, that I could
satisfy my own judgment even tolerably with the quality of
any article I produced; and my power to make sustained exertions drooped in a way I could not control, every other
hour of the day; insomuch that, what with parts to be can-
celled, and what with whole days of torpor and pure defect
of power to produce any thing at all, very often it turned out
that all my labors were barely sufficient (sometimes not sufficient) to meet the current expenses of my residence in London. Gloomy indeed was my state of mind at that period,
for though I made prodigious efforts to recover my health,
yet all availed me not, and a curse seemed to settle upon
whatever I then undertook. One canopy of murky clouds
brooded forever upon my spirits, which were in one uniformly low key of cheerless despondency."

De Quincey has given his views pretty freely as to the
regimen to be observed by reforming opium-eaters, in a paper
on " The Temperance Movement" which is specially worthy
of attention.

" My own experience had never travelled in that course
which could much instruct me in the miseries from wine or

in the resources for struggling with it. I had repeatedly been obliged, indeed, to lay it aside altogether ; but in this I never found room for more than seven or ten days' struggle : excesses I had never practiced in the use of wine : simply the habit of using it, and the collateral habits formed by excessive use of opium, had produced no difficulty at all in resigning it even on an hour's notice. From opium I derive my right of offering hints at all upon the subject of abstinence in other forms. But the modes of suffering from the evil, and the separate modes of suffering from the effort of self-conquest, together with errors of judgment incident to such states of transitional torment, are all nearly allied, practically analogous as regards the remedies, even if characteristically distinguished to the inner consciousness. I make no scruple, therefore, of speaking as from a station of high experience and of most watchful attention, which never remitted even under sufferings that were at times absolutely frantic. Once for all, however, in cases deeply rooted no advances ought ever to be made but by small stages ; for the effect, which is insensible at first, by the tenth, twelfth, or fifteenth day generally accumulates unendurably under any bolder deduction. Certain it is, that by an error of this nature at the outset, most natural to human impatience under exquisite suffering, too generally the trial is abruptly brought to an end through the crisis of a passionate relapse.

" Another object, and one to which the gladiator matched in single duel with intemperance must direct a religious vigilance, is the digestibility of his food. It must be digestible not only by its original qualities, but also by its culinary preparation.

" The whole process and elaborate machinery of digestion are felt to be mean and humiliating when viewed in relation to our mere animal economy. But they rise into dignity and assert their own supreme importance when they are studied from another station, viz., in relation to the intellect and temper. No man dares *then* to despise them ; it is then seen that these functions of the human system form the essential basis upon which the strength and health of our higher nature repose ; and that upon these functions, chiefly, the general happiness of life is dependent. All the rules of prudence or gifts of experience that life can accumulate, will never do as much for human comfort and welfare as would be done by a stricter attention, and a wiser science, directed to the digestive system. In this attention lies the key to any perfect restoration for the victim of intemperance. The sheet-anchor for the storm-beaten sufferer who is laboring to recover a haven of rest from the agonies of intemperance, and who has had the fortitude to abjure the poison which ruined, but which also for brief intervals offered him his only consolation, lies, beyond all doubt, in a most anxious regard to every thing connected with this supreme function of our animal economy. By how much the organs of digestion are feebler, by so much is it the more indispensable that solid and animal food should be adopted. A robust stomach may be equal to the trying task of supporting a fluid such as tea for breakfast ; but for a feeble stomach, and still worse for a stomach *enfeebled* by bad habits, broiled beef or something equally solid and animal, but not too much subjected to the action of fire, is the only tolerable diet. This indeed is the capital rule for a sufferer from habitual intoxication, who must inevitably labor under an impaired digestion : that as little as possible he should use of any liquid diet, and as little as possible of vegetable diet. Beef and a little bread (at the least sixty hours old) compose the privileged bill of fare for his breakfast. Errors of digestion, either from impaired powers or from powers not so much enfeebled as deranged, is the one immeasurable source both of disease and of secret wretchedness to the human race. Next, after the most vigorous attention, and a scientific attention, to the digestive system, in power of operation, stands *exercise*. For myself, under the ravages of opium, I have found walking the most beneficial exercise; besides that, it requires no previous notice or preparation of any kind; and this is a capital advantage in a state of drooping energies, or of impatient and unresting agitation. I may mention, as possibly an accident of my individual temperament, but possibly, also, no accident at all, that the relief obtained by walking was always most sensibly brought home to my consciousness, when some part of it (at least a mile and a half) had been performed before breakfast. In this there soon ceased to be any difficulty; for, while under the full oppression of opium it was impossible for me to rise at any hour that could, by the most indulgent courtesy, be described as within the pale of morning, no sooner had there been established any considerable relief from this oppression than the tendency was in the opposite direction—the difficulty became continually greater of sleeping even to a reasonable hour. Having once accomplished the feat of walking at 9 A.M., I backed in a space of seven or eight months to eight o'clock, to seven, to six, five, four, three; until at this point a metaphysical fear fell upon me that I was actually backing into 'yesterday,' and should soon have no sleep at all. Below three, however, I did not descend; and, for a couple of years, three and a half hours' sleep was all that I could obtain in the twenty-four hours. From this no particular suffering arose, except the nervous impatience of lying in bed for one moment after awaking. Consequently the habit of walking before breakfast became at length troublesome no longer as a most odious duty, but on the contrary, as a temptation that could hardly be resisted on the wettest mornings. As to the quantity of the exercise, I found that six miles a day formed the *minimum* which would support permanently a particular standard of animal spirits, evidenced to myself by certain apparent symptoms. I averaged about nine and a half miles a day, but ascended on particular days to fifteen or sixteen, and more rarely to twenty-three or twenty-four ; a quantity which did not produce fatigue : on the contrary it spread a sense of improvement through almost the whole week that followed ; but usually, in the night immediately succeeding to such an exertion, I lost much of my sleep—a privation that under the circumstances explained, deterred me from trying the experiment too often. For one or two years I accomplished more than I have here claimed, viz., from six to seven thousand miles in the twelve months.

" A necessity more painful to me by far than that of taking continued exercise arose out of a cause which applies perhaps with the same intensity only to opium cases, but must also apply in some degree to all cases of debilitation from morbid stimulation of the nerves, whether by means of wine, or opium, or distilled liquors. In travelling on the outside of mails during my youthful days, I made the discovery that opium, after an hour or so, diffuses a warmth deeper and far more permanent than could be had from any other known source. I mention this to explain in some measure the awful passion of cold which for some years haunted the inverse process of laying aside the opium. It was a perfect frenzy of misery ; cold was a sensation which then first, as a mode of torment, seemed to have been revealed. In the months of July and August, and not at all the less during the very middle watch of the day, I sat in the closest proximity to a blazing fire : cloaks, blankets, counterpanes, hearth-rugs, horse-cloths, were piled upon my shoulders, but with hardly a glimmering of relief.

"At night, and after taking coffee, I felt a little warmer, and could sometimes afford to smile at the resemblance of my own case to that of Harry Gile. Meantime, the external phenomenon by which the cold expressed itself was a sense (but with little reality) of eternal freezing perspiration. From this I was never free ; and at length, from finding one general ablution sufficient for one day, I was thrown upon the irritating necessity of repeating it more frequently than would seem credible if stated. At this time I used always hot water, and a thought occurred to me very seriously that it would be best to live constantly, and perhaps to sleep, in a bath. What caused me to renounce this plan was an accident that compelled me for one day to use cold water. This, first of all, communicated any lasting warmth ; so that ever afterward I used none *but* cold water. Now to live in a cold bath in our climate, and in my own state of preternatural sensibility to cold, was not an idea to dally with. I wish to mention, however, for the information of other sufferers in the same way, one change in the mode of applying the water which led to a considerable and a sudden improvement in the condition of my feelings. I had endeavored in vain to procure a child's battledore, as an easy means (when clothed with sponge) of reaching the interspace between the shoulders. In default of a battledore, therefore, my necessity threw my experiment upon a long hair-brush ; and this, eventually, proved of much greater service than any sponge or any battledore, for the friction of the brush caused an irritation on the surface of the skin, which, more than any thing else, has gradually diminished the once continual misery of unrelenting frost, although even yet it renews itself most distressingly at uncertain intervals.

" I counsel the patient not to make the mistake of supposing that his amendment will necessarily proceed continuously or by equal increments, because this, which is a common
notion, will certainly lead to dangerous disappointments.
How frequently I have heard people encouraging a self-re-
former by such language as this : ' When you have got
over the fourth day of abstinence, which suppose to be Sunday, then Monday will find you a trifle better ; Tuesday better still—though still it should be only a trifle—and so on.
You may at least rely on never going back, you may assure
yourself of having seen the worst, and the positive improve-
ments, if trifles separately, must soon gather into a sensible
magnitude.' This may be true in a case of short standing,
but as a general rule it is perilously delusive. On the contrary, the line of progress, if exhibited in a geometrical construction, would describe an ascending path upon the whole,
but with frequent retrocessions into descending curves, which,
compared with the point of ascent that had been previously
gained and so vexatiously interrupted, would sometimes seem
deeper than the original point of starting. This mortifying
tendency I can report from experience, many times repeated,
with regard to opium, and so unaccountably, as regarded all
the previous grounds of expectation, that I am compelled to
suppose it a tendency inherent in the very nature of all self-
restorations for animal systems.

" I counsel the patient frequently to call back before his
thoughts—when suffering sorrowful collapses that seem un-
merited by any thing done or neglected—that such, and far
worse perhaps, must have been his experience, and with no
reversion of hope behind, had he persisted in his intemperate indulgences ; *these* also suffer their own collapses, and
(so far as things not co-present can be compared) by many
degrees more shocking to the genial instincts. I exhort

him to believe that no movement on his own part, not
the smallest conceivable, toward the restoration of his
healthy state, can by possibility perish. Nothing in this
direction is finally lost; but often it disappears and hides
itself; suddenly, however, to re-appear, and in unexpected
strength, and much more hopefully, because such minute
elements of improvement, by re-appearing at a remoter stage,
show themselves to have combined with other elements of
the same kind, so that equally by their gathering tendency
and their duration through intervals of apparent darkness,
and below the current of what seemed absolute interruption,
they argue themselves to be settled in the system. There is
no good gift that does not come from God. Almost his greatest is health, with the peace which it inherits, and man must
reap *this* on the same terms as he was told to reap God's
earliest gift, the fruits of the earth, viz., 'in the sweat of his
brow,' through labor, often through sorrow, through disappointment, but still through imperishable perseverance, and
hoping under clouds when all hope seemed darkened.

" But it seems to me important not to omit this particular
caution : The patient will be naturally anxious, as he goes
on, frequently to test the amount of his advance, and its rate,
if that were possible ; but this he will see no mode of doing
except through tentative balancings of his feelings, and
generally of the moral atmosphere around him, as to pleasure
and hope, against the corresponding states so far as he can
recall them from his periods of intemperance. But these
comparisons I warn him are fallacious when made in this
way. The two states are incommensurable on any plan of
direct comparison. Some common measure must be found,
and *out of himself;* some positive fact that will not bend to
his own delusive feeling at the moment ; as, for instance, in
what degree he finds tolerable what heretofore was *not* so—

the effort of writing letters, or transacting business, or undertaking a journey, or overtaking the arrears of labor, that had been once thrown off to a distance. If in these things he finds himself improved, by tests that can not be disputed, he may safely disregard any sceptical whispers from a wayward sensibility which can not yet, perhaps, have recovered its normal health, however much improved. His inner feelings may not yet point steadily to the truth, though they may vibrate in that direction. Besides, it is certain that sometimes very manifest advances, such as any medical man would perceive at a glance, carry a man through stages of agitation and discomfort. A far worse condition might happen to be less agitated, and so far more bearable. Now when a man is positively suffering discomfort, when he is below the line of pleasurable feeling, he is no proper judge of his own condition, which he neither will nor can appreciate. Toothache extorts more groans than dropsy."

Little is definitely known to the public of De Quincey's opium habits subsequent to the publication in the year 1822 of the Appendix to the "Confessions." In the "Life of Professor Wilson," by his daughter, Mrs. Gordon, a letter from De Quincey, under date of February, 1824, is given, which says : "As to myself—though I have written not as one who labors under much depression of mind—the fact is, I *do* so. At this time calamity presses upon me with a heavy hand. I am quite free of opium, but it has left the liver, which is the Achilles heel of almost every human fabric, subject to affections which are tremendous for the weight of wretchedness attached to them. To fence with these with the one hand, and with the other to maintain the war with the wretched business of hack author, with all its horrible degradations, is more than I am able to bear. At this moment I have not a

place to hide my head in. Something I meditate—I know
not what—'*Itaque e conspectu omnium abiit.*' With a good pub-
lisher and leisure to premeditate what I write, I might yet
liberate myself; after which, having paid every body, I would
slink into some dark corner, educate my children, and show
my face in the world no more." To the statement of De
Quincey that he was then free of opium, Mrs. Gordon adds
in a note : " To the very last he asserted this, but the habit,
although modified, was never abandoned." Referring to a
protracted visit made by him in the year 1829-30 to Professor Wilson, Mrs. Gordon says :

" His tastes were very simple, though a little troublesome,
at least to the servant who prepared his repast. Coffee,
boiled rice and milk, and a piece of mutton from the loin
were the materials that invariably formed his diet. The
cook, who had an audience with him daily, received her instructions in silent awe, quite overpowered by his manner,
for had he been addressing a duchess he could scarcely
have spoken with more deference. He would couch his request in such terms as these : ' Owing to dyspepsia affecting
my system, and the possibility of any additional disarrange-
ment of the stomach taking place, consequences incalculably
distressing would arise, so much so indeed as to increase
nervous irritation, and prevent me from attending to matters
of overwhelming importance, if you do not remember to cut
the mutton in a diagonal rather than in a longitudinal form.'
But these little meals were not the only indulgences that,
when not properly attended to, brought trouble to Mr. De
Quincey. Regularity in doses of opium was even of greater
consequence. An ounce of laudanum per diem prostrated
animal life in the early part of the day. It was no unfre-
quent sight to find him in his room, lying upon the rug in
front of the fire, his head resting upon a book, his arms

crossed over his breast, plunged in profound slumber. For several hours he would lie in this state, until the effects of the torpor had passed away. The time when he was most brilliant was generally toward the early morning hours ; and then, more than once, in order to show him off, my father arranged his supper-parties so that, sitting till three or four in the morning, he brought Mr. De Quincey to that point at which in charm and power of conversation he was so truly wonderful."

In the "Suspiris de Profundis" of De Quincey, written in the year 1845, we have his own final record of the last chapter of his opium history. He says :

"In 1821, as a contribution to a periodical work—in 1822, as a separate volume—appeared the 'Confessions of an English Opium-Eater.' At the close of this little work the reader was instructed to believe, and *truly* instructed, that I had mastered the tyranny of opium. The fact is, that *twice* I mastered it, and by efforts even more prodigious in the second of these cases than in the first. But one error I committed in both. I did not connect with the abstinence from opium, so trying to the fortitude under *any* circumstances, that enormity of exercise which (as I have since learned) is the one sole resource for making it endurable. I overlooked, in those days, the one *sine qua non* for making the triumph permanent. Twice I sank, twice I rose again. A third time I sank ; partly from the cause mentioned (the oversight as to exercise), partly from other causes, on which it avails not now to trouble the reader. I could moralize if I chose ; and perhaps *he* will moralize whether I choose it or not. But in the mean time neither of us is acquainted properly with the circumstances of the case ; I, from natural bias of judgment, not altogether acquainted ; and he (with his permission) not at all.

" During this third prostration before the dark idol, and after some years, new and monstrous phenomena began slowly to arise. For a time these were neglected as accidents, or palliated by such remedies as I knew of. But when I could no longer conceal from myself that these dreadful symptoms were moving forward forever, by a pace steadily, solemnly, and equably increasing, I endeavored, with some feeling of panic, for a third time to retrace my steps. But I had not reversed my motions for many weeks before I became profoundly aware that this was impossible. Or, in the imagery of my dreams, which translated every thing into their own language, I saw through vast avenues of gloom those towering gates of ingress, which hitherto had always seemed to stand open, now at last barred against my retreat, and hung with funeral crape.

" The sentiment which attends the sudden revelation that *all is lost!* silently is gathered up into the heart; it is too deep for gestures or for words ; and no part of it passes to the outside. Were the ruin conditional, or were it in any point doubtful, it would be natural to utter ejaculations, and to seek sympathy. But where the ruin is understood to be absolute, where sympathy can not be consolation, and counsel can not be hope, this is otherwise. The voice perishes ; the gestures are frozen ; and the spirit of man flies back upon its own centre. I, at least, upon seeing those awful gates closed and hung with draperies of woe, as for a death already past, spoke not, nor started, nor groaned. One profound sigh ascended from my heart, and I was silent for days."*

* Mr. De Quincey died at Edinburgh, Dec. 8, 1859.

Soon after the death of Samuel Taylor Coleridge, a retired book-seller of Bristol by the name of Joseph Cottle
felt called upon to make public what he knew or could
gather respecting the opium habits of the philosopher and
poet. His first publication was made in the year 1837, and
was entitled " Recollections of Coleridge." Ten years later
he elaborated this publication into " The Reminiscences of
Coleridge and Southey." From the pages of the latter, from
Gilman's "Life of Coleridge," from the poet's own correspondence, and from the miscellaneous writings of De Quincey,
the following record has been chiefly compiled. From these
sources the reader can obtain a pretty accurate knowledge of
the circumstances under which Coleridge became an opium-
eater ; of the struggles he made to emancipate himself from
the habit, and of the intellectual ruin which opium entailed
upon one of the most marvellous-minded men the world has
produced.

It seems certain that Coleridge became familiar with opium as early at least as the year 1796, though it is probable
that its use did not become habitual till about 1802 or 1803.
From this period to the year 1814, his consumption of laudanum appears to have been enormous. The efforts he made
at self-reformation immediately previous to his admission in
1816 into the family of Dr. Gilman, were unsuccessful ; and
while the quantity of laudanum to which he had been so

long accustomed, was subsequently reduced to a small daily allowance, the opium *habit* ceased only with his life.

In justice to his memory, and in part mitigation of the censures of many of his personal friends, as well as to enable the reader to judge of the circumstances under which this distinguished man fell into his ruinous habit, the following extracts from his own letters and from other sources are given, nearly in chronological order, that it may be seen how far, from his childhood to his grave, Coleridge's constitutional infirmities furnish a partial apology for his excesses. Under date of Nov. 5, 1796, he writes to a friend :

" I wanted such a letter as yours, for I am very unwell. On Wednesday night I was seized with an intolerable pain from my temple to the tip of my right shoulder, including my right eye, cheek, jaw, and that side of the throat. I was nearly frantic, and ran about the house almost naked, endeavoring by every means to excite sensation in different parts of my body, and so to weaken the enemy by creating a diversion. It continued from one in the morning till half-past five, and left me pale and faint. It came on fitfully, but not so violently, several times on Thursday, and began severer threats toward night ; but I took between sixty and seventy drops of laudanum, and sopped the Cerberus just as his mouth began to open. On Friday it only niggled, as if the chief had departed, as from a conquered place, and merely left a small garrison behind, or as if he had evacuated the Corsica, and a few straggling pains only remained. But this morning he returned in full force, and his name is Legion. Giant-fiend of a hundred hands, with a shower of arrowy death-pangs he transpierced me ; and then he became a Wolf, and lay gnawing my bones ! I am not mad, most noble Festus ! but in sober sadness I have suffered this day more bodily pain than I had before a conception of. My right cheek has certainly been placed with admirable exactness under the focus of some invisible burning-glass, which concentrated all the rays of a Tartarean sun. My medical attendant decides it to be altogether nervous, and that ♠ originates either in severe application or excessive anxiety. My beloved Poole, in excessive anxiety I believe it might originate. I have a blister under my right ear, and I take twenty-five drops of laudanum every five hours, the ease and spirits gained by which have enabled me to write to you this flighty but not exaggerating account."

About the same time he writes to another friend, " A devil, a very devil, has got possession of my left temple, eye, cheek, jaw, throat, and shoulder. I can not see you this evening. I write in agony." Frequent reference is made in Coleridge's correspondence to his sufferings, from rheumatic or neuralgic affections, and the following letter, written in 1797, may possibly explain their origin :

" I had asked my mother one evening to cut my cheese entire, so that I might toast it. This was no easy matter, it being a *crumbly* cheese. My mother, however, did it. I went into the garden for something or other, and in the mean time my brother Frank minced my cheese, to ' disappoint the favorite.' I returned, saw the exploit, and in an agony of passion flew at Frank. He pretended to have been seriously hurt by my blow, flung himself on the ground, and there lay with outstretched limbs. I hung over him mourning and in a great fright ; he leaped up, and with a horse-laugh gave me a severe blow in the face. I seized a knife and was running at him, when my mother came in and took me by the arm. I expected a flogging, and struggling from her I ran away to a little hill or slope, at the bottom of which the Otter flows, about a mile from Ottery. There I stayed. My rage died away, but my obstinancy vanquished my fears, and taking

out a shilling book, which had at the end morning and evening prayers, I very devoutly repeated them—thinking at the
same time, with a gloomy inward satisfaction, how miserable
my mother must be ! . . . . It grew dark and I fell asleep.
It was toward the end of October, and it proved a stormy
night. I felt the cold in my sleep, and dreamed that I was
pulling the blanket over me, and actually pulled over me a
dry thorn-bush which lay on the ground near me. In my
sleep I had rolled from the top of the hill till within three
yards of the river, which flowed by the unfenced edge of
the bottom. I awoke several times, and finding myself wet,
and cold, and stiff, closed my eyes again that I might forget it.

"In the mean time my mother waited about half an hour,
expecting my return when the *sulks* had evaporated. I not
returning, she sent into the church-yard and round the town.
Not found! Several men and all the boys were sent out to
ramble about and seek me. In vain! My mother was almost distracted, and at ten o'clock at night I was *cried* by
the crier in Ottery and in two villages near it, with a reward
offered for me. No one went to bed ; indeed I believe half
the town were up all the night. To return to myself. About
five in the morning, or a little after, I was broad awake and
attempted to get up and walk, but I could not move. I saw
the shepherds and workmen at a distance and cried, but so
faintly that it was impossible to hear me thirty yards off, and
there I might have lain and died—for I was now almost given
over, the pond and even the river near which I was lying
having been dragged—but providentially Sir Stafford Northcote, who had been out all night, resolved to make one other
trial, and came so near that he heard me crying. He carried
me in his arms for nearly a quarter of a mile, when we met
my father and Sir Stafford Northcote's servants. I remem-

blessed him with a buoyancy of spirits, and even when suffering he deceived the partial observer.

" At this time (while a soldier) he frequently complained
of a pain at the pit of his stomach, accompanied with sickness, which totally prevented his stooping, and in consequence
he could never arrive at the power of bending his body to
rub the heels of his horse.   During the latter part of his life
he became nearly crippled by the rheumatism."

Under date of July 24, 1800, Coleridge writes : " I have
been more unwell than I have ever been since I left school.
For many days I was forced to keep my bed, and when released from that incarceration I suffered most grievously
from a brace of swollen eyelids and a head into which, on
the least agitation, the blood was felt as rushing in and flowing back again, like the raking of the tide on a coast of loose
stones."

In January, 1803, he says : " I write with difficulty, with
all the fingers but one of my right hand very much swollen.
Before I was half up the *Kirkstone* mountain, the storm had
wetted me through and through.   In spite of the wet and
the cold I should have had some pleasure in it, but for two
vexations ; first, an almost intolerable pain came into my
right eye, a smarting and burning pain ; and secondly, in
consequence of riding with such cold water under my seat,
extremely uneasy and burdensome feelings attacked my
groin, so that, what with the pain from the one, and the
alarm from the other, I had no enjoyment at all !

" I went on to Grasmere.   I was not at all unwell when
I arrived there, though wet of course to the skin.   My right
eye had nothing the matter with it, either to the sight of
others or to my own feelings, but I had a bad night with
distressful dreams, chiefly about my eye ; and waking often
in the dark, I thought it was the effect of mere recollection,

-

but it appeared in the morning that my right eye was bloodshot and the lid swollen. That morning, however, I walked home, and before I reached Keswick my eye was quite well, but *I felt unwell all over.* Yesterday I continued unusually unwell all over me till eight o'clock in the evening. I took no *laudanum or opium*, but at eight o'clock, unable to bear the stomach uneasiness and aching of my limbs, I took two large tea-spoons full of ether in a wine-glass of camphorated gum-water, and a third tea-spoon full at ten o'clock, and I received complete relief, my body calmed, my sleep placid; but when I awoke in the morning my right hand, with three of the fingers, were swollen and inflamed. The swelling in the hand is gone down, and of two of the fingers somewhat abated, but the middle finger is still twice its natural size, so that I write with difficulty."

A few days later, he writes to the same friend: "On Monday night I had an attack in my stomach and right side, which in pain, and the length of its continuance, appeared to me by far the severest I ever had. About one o'clock the pain passed out of my stomach, like lightning from a cloud, into the extremities of my right foot. My toe swelled and throbbed, and I was in a state of delicious ease which the pain in my toe did not seem at all to interfere with. On Wednesday I was well, and after dinner wrapped myself up warm and walked to Lodore.

"The walk appears to have done me good, but I had a wretched night: shocking pains in my head, occiput, and teeth, and found in the morning that I had two bloodshot eyes. But almost immediately after the receipt and perusal of your letter the pains left me, and I am bettered to this hour; and am now indeed as well as usual saving that my left eye is very much bloodshot. It is a sort of duty with me to be particular respecting parts that relate to my health. I have retained a good sound appetite through the whole of it, without any craving after exhilarants or narcotics; and I have got well as in a moment. Rapid recovery is constitutional with me ; but the former circumstances I can with certainty refer to the system of diet, abstinence of vegetables, wine, spirits, and beer, which I have adopted by your advice."

The same year he writes to a friend suffering from a chronic disorder, and records the trial of Bang—"the powder of the leaves of a kind of hemp that grows in the hot climates. It is prepared, and I believe used, in all parts of the east, from Morocco to China. In Europe it is found to act very differently on different constitutions. Some it elevates in the extreme ; others it renders torpid, and scarcely observant of any evil that may befall them. In Barbary it is always taken, if it can be procured, by criminals condemned to suffer amputation, and it is said to enable those miserables to bear the rough operations of an unfeeling executioner more than we Europeans can the keen knife of our most skillful chirurgeons :

"We will have a fair trial of Bang. Do bring down some of the Hyoscyamine pills, and I will give a fair trial to Opium, Henbane, and Nepenthe. By the bye, I always considered Homer's account of the Nepenthe as a *Banging* lie."

In September, 1803, he gives a gloomy account of his condition. It seems probable that at this time his use of opium must have become habitual :

"For five months past my mind has been strangely shut up. I have taken the paper with the intention to write to you many times, but it has been one blank feeling—one blank idealess feeling. I had nothing to say—could say nothing. How dearly I love you, my very dreams make known to me. I will not trouble you with the gloomy tale of my health. When I am awake, by patience, employment, effort of mind, and walking, I can keep the fiend at arm's-length, but the night is my Hell! sleep my tormenting Angel. Three nights out of four I fall asleep, struggling to lie awake, and my frequent night-screams have almost made me a nuisance in my own house. Dreams with me are no shadows, but the very calamities of my life.

"In the hope of drawing the gout, if gout it should be, into my feet, I walked, previously to my getting into the coach at Perth, 263 miles in eight days, with no unpleasant fatigue. My head is equally strong ; but acid or not acid, gout or not gout, something there is in my stomach.

"To diversify this dusky letter, I will write an *Epitaph*, which I composed in my sleep for myself while dreaming that I was dying. To the best of my recollection I have not altered a word :

> "'Here sleeps at length poor Col. and without screaming,
> Who died as he had always lived, a dreaming ;
> Shot dead, while sleeping, by the gout within,
> Alone, and all unknown, at E'nbro' in an Inn.'"

In the beginning of the next year, 1804, the state of his health is thus indicated : "I stayed at Grasmere (Mr. Wordsworth's) a month—three-fourths of the time bedridden—and deeply do I feel the enthusiastic kindness of Wordsworth's wife and sister, who sat up by me, one or the other, in order to awaken me at the first symptoms of distressful feeling ; and even when they went to rest, continued often and often to weep and watch for me even in their dreams.

"Though my right hand is so much swollen that I can scarcely keep my pen steady between my thumb and finger, yet my stomach is easy and my breathing comfortable, and I am eager to hope all good things of my health. That gained, I have a cheering and I trust prideless confidence that I shall make an active and perseverant use of the faculties and requirements that have been entrusted to my keeping, and a fair trial of their height, depth, and width."

A few days later he writes to a friend who was suffering like himself : " Have you ever thought of trying large doses of opium, a hot climate, keeping your body open by grapes, and the fruits of the climate ? Is it possible that by drinking freely you might at last produce the gout, and that a violent pain and inflammation in the extremities might produce new trains of motion and feeling in your stomach, and the organs connected with the stomach, known and unknown ? I know by a little what your sufferings are, and that to shut the eyes and stop up the ears is to give one's self up to storm and darkness, and the lurid forms and horrors of a dream."

In reference to these statements regarding Coleridge's physical condition, Cottle remarks : "I can testify that, during the four or five years in which Mr. C. resided in or near Bristol, no young man could enjoy more robust health. Dr. Carlyon also verbally stated that Mr. C., both at Cambridge and at Gottingen, 'possessed sound health.' From these premises the conclusion is fair that Mr. Coleridge's unhappy use of narcotics, which commenced thus early, was the true cause of all his maladies, his languor, his acute and chronic pains, his indigestion, his swellings, the disturbances of his general corporeal system, his sleepless nights, and his terrific dreams."

Scattered through Dr. Gilman's "Life of Coleridge" are indications of this kind :

"In 1804, his rheumatic sufferings increasing, he determined on a change of climate, and went in May to Malta. He seemed at this time, in addition to his rheumatism, to have been oppressed in his breathing, which oppression crept on him, imperceptibly to himself, without suspicion of its

G

cause. Yet so obvious was it that it was noticed by others 'as laborious;' and continuing to increase, though with little apparent advancement, at length terminated in death.

"At first he remarked that he was relieved by the climate of Malta, but afterward speaks of his limbs 'as lifeless tools,' and of the violent pain in his bowels, which neither opium, ether, nor peppermint, separately or combined, could relieve.

"Coleridge *began* the use of opium from bodily pain (rheumatism), and for the same reason *continued* it, till he had acquired a habit too difficult under his own management to control. To him it was the thorn in the flesh, which will be seen in the following note found in his pocket-book: 'I have never loved evil for its own sake; no! nor ever sought pleasure for its own sake, but only as the means of escaping from pains that coiled around my mental powers as a serpent around the body and wings of an eagle! My sole sensuality was *not* to be in pain.'"

Little is known of Coleridge's opium habits during his residence at Malta. On his return to England in 1807, he wrote to Mr. Cottle: "On my return to Bristol, whenever that may be, I will certainly give you the right hand of old fellowship; but, alas! you will find me the wretched wreck of what you knew me, rolling, rudderless. My health is extremely bad. Pain I have enough of, but that is indeed to me a mere trifle, but the almost unceasing, overpowering sensations of wretchedness—achings in my limbs, with an indescribable restlessness that makes action to any available purpose almost impossible—and worst of all the sense of blighted utility, regrets, not remorseless. But enough; yea, more than enough, if these things produce or deepen the conviction of the utter powerlessness of ourselves, and that we either perish or find aid from something that passes understanding."

A period of seven years here intervenes, during which no

light is thrown upon the opium life of Coleridge. The following extract from a letter written by him during this period, sufficiently indicates, however, both his consciousness of his great powers and his remorse for their imperfect use:

"As to the letter you propose to write to a man who is unworthy even of a rebuke from you, I might most unfeignedly object to some parts of it from a pang of conscience forbidding me to allow, even from a dear friend, words of admiration which are inapplicable in exact proportion to the power given to me of having deserved them if I had done my duty.

"It is not of comparative utility I speak; for as to what has been actually done, and in relation to useful effects produced—whether on the minds of individuals or of the public—I dare boldly stand forward, and (let every man have his own, and that be counted mine which but for and through me would not have existed) will challenge the proudest of my literary contemporaries to compare proofs with me of usefulness in the excitement of reflection, and the diffusion of original or forgotten yet necessary and important truths and knowledge; and this is not the less true because I have suffered others to reap all the advantages. But, O dear friend, this consciousness, raised by insult of enemies and alienated friends, stands me in little stead to my own soul—in how little, then, before the all-righteous Judge! who, requiring back the talents he had entrusted, will, if the mercies of Christ do not intervene, not demand of me what I have done, but why I did not do more; why, with powers above so many, I had sunk in many things below most!"

In 1814 he returned to Bristol, and here the painful narrative of Mr. Cottle comes in: "Is it expedient, is it lawful, to give publicity to Mr. Coleridge's practice of inordinately taking opium ; which to a certain extent, at one part of his life, inflicted on a heart naturally cheerful the stings of conscience, and sometimes almost the horrors of despair?

" In the year 1814, all this, I am afflicted to say, applied to Mr. Coleridge. Once Mr. Coleridge expressed to me, with indescribable emotion, the joy he should feel if he could collect around him all who were ' beginning to tamper with the lulling but fatal draught,' so that he might proclaim as with a trumpet, ' the worse than death that opium entailed.'

" When it is considered, also, how many men of high mental endowments have shrouded their lustre by a passion for this stimulus, would it not be a criminal concession to unauthorized feelings to allow so impressive an exhibition of this subtle species of intemperance to escape from public notice? In the exhibition here made, the inexperienced in future may learn a memorable lesson, and be taught to shrink from opium as they would from a scorpion, which, before it destroys, invariably expels peace from the mind, and excites the worst species of conflict—that of setting a man at war with himself.

" I had often spoken to Hannah More of S. T. Coleridge, and proceeded with him one morning to Barley Wood, her residence, eleven miles from Bristol. The interview was mutually agreeable, nor was there any lack of conversation ; but I was struck with something singular in Mr. Coleridge's eye. I expressed to a friend, the next day, my concern at having beheld him during his visit to Hannah More so extremely paralytic, his hands shaking to an alarming degree, so that he could not take a glass of wine without spilling it, though one hand supported the other ! ' That,' said he, ' arises from the immoderate quantity of OPIUM he takes.'

" It is remarkable that this was the first time the melancholy fact of Mr. Coleridge's excessive indulgence in opium had come to my knowledge. It astonished and afflicted me. Now the cause of his ailments became manifest. On this subject Mr. C. may have been communicative to others, but to me he was silent.

" I ruminated long upon this subject with indescribable sorrow ; and having ascertained from others not only the existence of the evil but its extent, I determined to write to Mr. Coleridge. I addressed him the following letter, under the full impression that it was a case of ' life and death,' and that if some strong effort were not made to arouse him from his insensibility, speedy destruction must inevitably follow.

" ' BRISTOL, April 25, 1814.

" ' DEAR COLERIDGE :—I am conscious of being influenced by the purest motives in addressing to you the following letter. Permit me to remind you that I am the oldest friend you have in Bristol, that I was such when my friendship was of more consequence to you than it is at present, and that at that time you were neither insensible of my kindnesses nor backward to acknowledge them. I bring these things to your remembrance to impress on your mind that it is still a *friend* who is writing to you ; one who ever has been such, and who is now going to give you the most decisive evidence of his sincerity.

" ' When I think of Coleridge I wish to recall the image of him such as he appeared in past years ; now, how has the baneful use of opium thrown a dark cloud over you and your prospects ! I would not say any thing needlessly harsh or unkind, but I must be *faithful.* It is the irresistible voice of conscience. Others may still flatter you, and hang upon your words, but I have another, though a less gracious duty to perform. I see a brother sinning a sin unto death, and shall I not warn him ? I see him perhaps on the borders of eternity ; in effect, despising his Maker's law, and yet indifferent to his perilous state !

" 'In recalling what the expectations concerning you once were, and the excellency with which seven years ago you wrote and spoke on religious truth, my heart bleeds to see how you are now fallen, and thus to notice how many exhilarating hopes are almost blasted by your present habits. This is said, not to wound, but to arouse you to reflection.

" 'I know full well the evidences of the pernicious drug ! You can not be unconscious of the effects, though you may wish to forget the cause. All around you behold the wild eye, the sallow countenance, the tottering step, the trembling hand, the disordered frame ! and yet will you not be awakened to a sense of your danger, and I must add, your guilt ? Is it a small thing, that one of the finest of human understandings should be lost ? That your talents should be buried ? That most of the influences to be derived from your present example should be in direct opposition to right and virtue ? It is true you still talk of religion, and profess the warmest admiration of the Church and her doctrines, in which it would not be lawful to doubt your sincerity ; but can you be unaware that by your unguarded and inconsistent conduct you are furnishing arguments to the infidel ; giving occasion for the enemy to blaspheme ; and (among those who imperfectly know you) throwing suspicion over your religious profession ? Is not the great test in some measure against you, " By their fruits ye shall know them ?" Are there never any calm moments, when you impartially judge of your own actions by their consequences ?

" 'Not to reflect on you—not to give you a moment's *needless* pain, but in the spirit of friendship, suffer me to bring to your recollection some of the sad effects of your undeniable intemperance.

"'I know you have a correct love of honest independence, without which there can be no true nobility of mind; and yet for opium you will sell this treasure, and expose yourself to the liability of arrest by some "dirty fellow" to whom you choose to be indebted for "ten pounds!" You had, and still have, an acute sense of moral right and wrong, but is not the feeling sometimes overpowered by self-indulgence? Permit me to remind you that you are not more suffering in your mind than you are in your body, while you are squandering largely your money in the purchase of opium, which, in the strictest equity, should receive a *different direction.*

"'I will not again refer to the mournful effects produced on your own health from this indulgence in opium, by which you have undermined your strong constitution; but I must notice the injurious consequences which this passion for the narcotic drug has on your literary efforts. What you have already done, excellent as it is, is considered by your friends and the world as the bloom, the mere promise of the harvest. Will you suffer the fatal draught, which is ever accompanied by sloth, to rob you of your fame, and, what to you is a higher motive, of your power of doing good; of giving fragrance to your memory, among the worthies of future years, when you are numbered with the dead?

"'And now let me conjure you, alike by the voice of friendship and the duty you owe yourself and family; above all, by the reverence you feel for the cause of Christianity; by the fear of God and the awfulness of eternity, to renounce from this moment opium and spirits as your bane! Frustrate not the great end of your existence. Exert the ample abilities which God has given you, as a faithful steward. So will you secure your rightful pre-eminence among the sons of genius; recover your cheerfulness, your health—I trust it is not too late—become reconciled to yourself; and, through the merits of that Saviour in whom you profess to trust, obtain at last the approbation of your Maker. My dear Coleridge, be wise before it be too late. I do hope to see you a renovated man; and that you will still burst your inglorious fetters and justify the best hopes of your friends.

"'Excuse the freedom with which I write. If at the first moment it should offend, on reflection you will approve at least of the motive, and perhaps, in a better state of mind, thank and bless me. If all the good which I have prayed for should not be effected by this letter, I have at least discharged an imperious sense of duty. I wish my manner were less exceptionable, as I do that the advice through the blessing of the Almighty might prove effectual. The tear which bedims my eye is an evidence of the sincerity with which I subscribe myself your affectionate friend,

"'JOSEPH CUTTLE.'

"The following is Mr. Coleridge's reply:

"'April 26, 1814.

"' You have poured oil in the raw and festering wound of an old friend's conscience Cottle, but it is *oil of vitriol!* I but barely glanced at the middle of the first page of your letter, and have seen no more of it—not from resentment, God forbid! but from the state of my bodily and mental sufferings, that scarcely permitted human fortitude to let in a · new visitor of affliction.

"' The object of my present reply is to state the case just as it is—first, that for ten years the anguish of my spirit has been indescribable, the sense of my danger staring, but the consciousness of my GUILT worse—far worse than all! I have prayed, with drops of agony on my brow; trembling not only before the justice of my Maker, but even before the mercy of my Redeemer. " I gave thee so many talents, what hast thou done with them ? " Secondly, overwhelmed as I am with a sense of my direful infirmity, I have never attempted to disguise or conceal the cause. On the contrary, not only to friends have I stated the whole case with tears and the very bitterness of shame, but in two instances I have warned young men—mere acquaintances, who had spoken of having taken laudanum—of the direful consequences, by an awful exposition of its tremendous effects on myself.

" ' Thirdly, though before God I can not lift up my eyelids, and only do not despair of his mercy because to despair would be adding crime to crime, yet to my fellow-men I may say that I was seduced into the ACCURSED habit ignorantly. I had been almost bedridden for many months with swellings in my knees. In a medical journal I unhappily met with an account of a cure performed in a similar case, or what appeared to me so, by rubbing in of laudanum, at the same time taking a given dose internally. It acted like a charm, like a miracle ! I recovered the use of my limbs, of my appetite, of my spirits, and this continued for near a fortnight. At length the unusual stimulus subsided, the complaint returned—the supposed remedy was recurred to— but I can not go through the dreary history.

" ' Suffice it to say that effects were produced which acted on me by terror and cowardice of pain and sudden death, not (so help me God !) by any temptation of pleasure, or expectation or desire of exciting pleasurable sensations. On the very contrary, Mrs. Morgan and her sister will bear witness so far as to say that the longer I abstained the higher my spirits were, the keener my enjoyments, till the moment, the direful moment arrived when my pulse began to fluctuate, my heart to palpitate, and such falling abroad as it were of my whole frame, such intolerable restlessness and

incipient bewilderment, that in the last of my several attempts to abandon the dire poison I exclaimed in agony,
which I now repeat in seriousness and solemnity, " I am too
poor to hazard this !"   Had I but a few hundred pounds—but
£200—half to send to Mrs. Coleridge, and half to place myself in a private mad-house, where I could procure nothing
but what a physician thought proper, and where a medical
attendant could be constantly with me for two or three
months (in less than that time life or death would be determined), then there might be hope.   Now there is none ! !   O
God ! how willingly would I place myself under Dr. Fox in
his establishment ; for my case is a species of madness, only
that it is a derangement, an utter impotence of the volition
and not of the intellectual faculties.   You bid me rouse myself.   Go bid a man paralytic in both arms to rub them brisk-
ly together and that will cure him.   " Alas !" he would reply,
"that I can not move my arms is my complaint and my misery."   May God bless you, and your affectionate but most
afflicted   ·   S. T. COLERIDGE.'

"On receiving this full and mournful disclosure I felt the
deepest compassion for Mr. C.'s state, and sent him a letter
to which I received the following reply :

" ' O, dear friend! I have too much to be forgiven to feel
any difficulty in forgiving the cruellest enemy that ever
trampled on me : and you I have only to *thank !* You have
no conception of the dreadful hell of my mind, and conscience, and body.   You bid me pray.   Oh, I do pray inward-
ly to be able to pray ; but indeed to pray, to pray with a faith
to which a blessing is promised, this is the reward of faith,
this is the gift of God to the elect.   Oh ! if to feel how infinitely worthless I am, how poor a wretch, with just free-will

enough to be deserving of wrath and of my own contempt, and of none to merit a moment's peace, can make a part of a Christian's creed—so far I am a Christian.

S. T. C.'

"'April 26, 1814.

"At this time Mr. Coleridge was indeed in a pitiable condition. His passion for opium had so completely subdued his *will* that he seemed carried away, without resistance, by an overwhelming flood. The impression was fixed on his mind that he should inevitably die unless he were placed under *constraint*, and that constraint he thought could be alone effected in an asylum. Dr. Fox, who presided over an establishment of this description in the neighborhood of Bristol, appeared to Mr. C. the individual to whose subjection he would most like to submit. This idea still impressing his imagination, he addressed to me the following letter :

"'DEAR COTTLE :—I have resolved to place myself in any situation in which I can remain for a month or two as a child, wholly in the power of others. But, alas! I have no money. Will you invite Mr. Hood, a most dear and affectionate friend to worthless me, and Mr. Le Breton, my old school-fellow and likewise a most affectionate friend, and Mr. Wade, who will return in a few days ; desire them to call on you, any evening after seven o'clock that they can make convenient, and consult with them whether any thing of this kind can be done. Do you know Dr. Fox? Affectionately,

"'S. T. C.'

"I *did* know the late Dr. Fox, who was an opulent and liberal-minded man, and if I had applied to him, or any friend had so done, I can not doubt but that he would instantly have received Mr. Coleridge gratuitously ; but nothing could have induced me to make the application but that extreme case which did not then appear fully to exist.

"The years 1814 and 1815 were the darkest periods in Mr. Coleridge's life. However painful the detail, it is presumed that the reader would desire a knowledge of the undisguised truth. This can not be obtained without introducing the following letters of Mr. Southey, received from him after having sent him copies of the letters which passed between Mr. Coleridge and myself.

"'KESWICK, April, 1814.

"'MY DEAR COTTLE:—You may imagine with what feelings I have read your correspondence with Coleridge. Shocking as his letters are, perhaps the most mournful thing they discover is, that while acknowledging the guilt of the habit he imputes it still to morbid bodily causes, whereas after every possible allowance is made for these, every person who has witnessed his habits knows that for the greater, infinitely the greater part, inclination and indulgence are its motives.

"'It seems dreadful to say this, with his expressions before me, but it is so, and I know it to be so from my own observation, and that of all with whom he has lived. The Morgans, with great difficulty and perseverance, *did* break him of the habit at a time when his ordinary consumption of laudanum was from *two quarts a week* to *a pint a day!* He suffered dreadfully during the first abstinence, so much so as to say it was better for him to die than to endure his present feelings. Mrs. Morgan resolutely replied, it was indeed better that he should die than that he should continue to live as he had been living. It angered him at the time, but the effort was persevered in.

"'To what, then, was the relapse owing? I believe to this cause—that no use was made of renewed health and spirits ;

that time passed on in idleness, till the lapse of time brought with it a sense of neglected duties, and then relief was again sought for *a self-accusing mind* in bodily feelings, which, when the stimulus ceased to act, added only to the load of self-accusation. This, Cottle, is an insanity which none but the soul's Physician can cure. Unquestionably, restraint would do as much for him as it did when the Morgans tried it, but I do not see the slightest reason for believing it would be more permanent. This, too, I ought to say, that all the medical men to whom Coleridge has made his confession have uniformly ascribed the evil not to bodily disease but indulgence. The restraint which alone could effectually cure is that which no person can impose upon him. Could he be compelled to a certain quantity of labor every day for his family, the pleasure of having done it would make his heart glad, and the sane mind would make the body whole.

" ' His great object should be to get out a play, and appropriate the whole produce to the support of his son Hartley at college. Three months' pleasurable exertion would effect this. Of some such fit of industry I by no means despair; of any thing more than fits I am afraid I do. But this of course I shall never say to him. From me he shall never hear aught but cheerful encouragement and the language of hope.'

" After anxious consideration I thought the only effectual way of benefiting Mr. Coleridge would be to renew the project of an annuity, by raising for him among his friends one hundred, or, if possible, one hundred and fifty pounds a year, purposing through a committee of three to pay for his comfortable board and all necessaries, but not of giving him the disposition of any part till it was hoped the correction of his bad habits and the establishment of his better principles might qualify him for receiving it for his own distribution. It was difficult to believe that his subjection to *opium* could much longer resist the stings of his own conscience and the solicitations of his friends, as well as the pecuniary destitution to whicĦ his *opium habits* had reduced him. The proposed object was named to Mr. C., who reluctantly gave his consent.

"I now drew up a letter, intending to send a copy to all Mr. Coleridge's old and steady friends (several of whom approved of the. design), but before any commencement was made I transmitted a copy of my proposed letter to Mr. Southey to obtain his sanction. The following is his reply :

"'April 17th, 1814.

"'DEAR COTTLE :—I have seldom in the course of my life felt it so difficult to answer a letter as on the present occasion. There is, however, no alternative. I must sincerely express what I think, and be thankful I am writing to one who knows me thoroughly.

"'Of sorrow and humiliation I will say nothing. No part of Coleridge's embarrassment arises from his wife and children, except that he has insured his life for a thousand pounds, and pays the annual premium. He never writes to them, and never opens a letter from them.

"'In truth, Cottle, his embarrassments and his miseries of body and mind all arise from one accursed cause—excess in *opium*, of which he habitually takes more than was ever known to be taken by any person before him. The Morgans, with great effort, succeeded in making him leave it off for a time, and he recovered in consequence *health* and *spirits*. He has now taken to it again. Of this indeed I was too sure before I heard from you—that his looks bore testimony to it.

Perhaps you are not aware of the costliness of this drug. In
the quantity which C. takes, it would consume *more* than the
whole which you propose to raise. A frightful consumption
of *spirits* is added. In this way bodily ailments are produced, and the wonder is that he is still alive.

" ' Nothing is wanting to make him easy in circumstances
and happy in himself but to leave off opium, and to direct a
certain portion of his time to the discharge of *his duties.*'

" During my illness at this time, Mr. Coleridge sent my
sister the following letter, and the succeeding one to myself :

" ' 13th May, 1814.

" ' DEAR MADAM :—I am uneasy to know how my friend,
J. Cottle, goes on. The walk I took last Monday to inquire
in person proved too much for my strength, and shortly after
my return I was in such a swooning way that I was directed
to go to bed, and orders were given that no one should inter-
rupt me. Indeed I can not be sufficiently grateful for the
skill with which *the surgeon treats me.* But it must be a slow,
and occasionally an interrupted progress, after a sad retro-
gress of nearly twelve years.'

" ' Friday, 27th May, 1814.

" ' MY DEAR COTTLE :—I feel, with an intensity unfathoma-
ble by words, my utter nothingness, impotence, and worth-
lessness, in and for myself. I have learned what a sin is
against an infinite, imperishable being, such as is the soul of
man.

" ' I have had more than a glimpse of what is meant by
death and outer darkness, and the worm that dieth not—and
that all the *hell* of the reprobate, is no more inconsistent with
the love of God, than the blindness of one who has occasioned loathsome and guilty diseases to eat out his eyes is

inconsistent with the light of the sun.   But the consolations,
at least the sensible sweetness of hope, I do not possess.   On
the contrary, the temptation which I have constantly to fight
up against, is a fear that if *annihilation* and the *possibility* of
heaven were offered to my choice, I should choose the former.

" ' Mr. Eden gave you a too flattering account of me.   It is
true I am restored, as much beyond my expectations almost
as my deserts ; but I am exceedingly weak.   I need for myself solace and refocillation of animal spirits, instead of being in a condition of offering it to others.'

" The serious expenditure of money resulting from Mr. C.'s
consumption of opium was the least evil, though very great,
and must have absorbed all the produce of Mr. C.'s lectures and all the liberalities of his friends.   It is painful to
record such circumstances as the following, but the picture
would be incomplete without it.

" Mr. Coleridge, in a late letter, with something it is feared,
if not of duplicity, of self-deception, extols the skill of his
surgeon in having gradually lessened his consumption of
laudanum, it was understood, to twenty drops a day.   With
this diminution the habit was considered as subdued, at
which result no one appeared to rejoice more than Mr. Coleridge himself.   The reader will be surprised to learn that, notwithstanding this flattering exterior, Mr. C., while apparently
submitting to the directions of his medical adviser, was secretly indulging in his usual overwhelming quantities of opium !
Heedless of his health and every honorable consideration, he
contrived to obtain surreptitiously the fatal drug, and thus to
baffle the hopes of his warmest friends.

" Mr. Coleridge had resided at this time for several months
with his kind friend Mr. Josiah Wade, of Bristol, who in his
solicitude for his benefit had procured for him, so long as it

was deemed necessary, the professional assistance stated above. The surgeon on taking leave, after the cure had been *effected*, well knowing the expedients to which opium patients would often recur to obtain their proscribed draughts—at least till the habit of temperance was fully established—cautioned Mr. W. to prevent Mr. Coleridge by all possible means from obtaining that by stealth from which he was openly debarred. It reflects great credit on Mr. Wade's humanity that, to prevent all access to opium, and thus if possible to rescue his friend from destruction, he engaged a respectable old decayed tradesman constantly to attend Mr. C., and, to make that which was sure, doubly certain, placed him even in his bedroom; and this man always accompanied him whenever he went out. To such surveillance Mr. Coleridge cheerfully acceded, in order to show the promptitude with which he seconded the efforts of his friends. It has been stated that every precaution was unavailing. By some unknown means and dexterous contrivances Mr. C. afterward confessed that he still obtained his usual lulling potions.

" As an example, among others of a similar nature, one ingenious expedient to which he resorted to cheat the doctor he thus disclosed to Mr. Wade, from whom I received it. He said, in passing along the quay where the ships were moored, he noticed by a side glance a druggist's shop, probably an old resort, and standing near the door he looked toward the ships, and pointing to one at some distance he said to his attendant, ' I think that's an American.' ' Oh, no, that I am sure it is not,' said the man. ' I think it is,' replied Mr. C. ' I wish you would step over and ask, and bring me the particulars.' The man accordingly went; when as soon as his back was turned Mr. C. stepped into the shop, had his portly bottle filled with laudanum, which he always carried in his pocket, and then expeditiously placed himself in the spot where he was left. The man now returned with the particulars, beginning, ' I told you, Sir, it was not an American, but I have learned all about her.' 'As I am mistaken, never mind the rest,' said Mr. C., and walked on.

"A common impression prevailed on the minds of his friends that it was a desperate case that paralyzed all their efforts ; that to assist Mr. C. with money, which under favorable circumstances would have been most promptly advanced, would now only enlarge his capacity to obtain the opium which was consuming him. We at length learned that Mr. Coleridge was gone to reside with his friend Mr. John Morgan, in a small house, at Calne, in Wiltshire. So gloomy were our apprehensions, that even the death of Mr. C. was mournfully expected at no distant period, for his actions at this time were, we feared, all indirectly of a suicidal description.

"In a letter dated October 27, 1814, Mr. Southey thus writes :

"'Can you tell me any thing of Coleridge? We know that he is with the Morgans at Calne. What is to become of him ? He may find men who will give him board and lodging for the sake of his conversation, but who will pay his other expenses ? He leaves his family to chance and charity. With good feelings, good principles, as far as the understanding is concerned, and an intellect as clear and as powerful as was ever vouchsafed to man, he is the slave of degrading sensuality, and sacrifices every thing to it. The case is equally deplorable and monstrous.'"

The intimacy between Coleridge and Cottle seems about this period to have entirely ceased. After the death of Coleridge, Mr. Cottle prepared his "Recollections" of his friend, but was restrained from its publication by considerations of propriety, until the following letter was placed in his hands by the gentleman to whom it was addressed, with permission to use it:

"BRISTOL, June 26, 1814.

"DEAR SIR :—For I am unworthy to call any good man friend — much less you, whose hospitality and love I have abused ; accept, however, my entreaties for your forgiveness and your prayers.

"Conceive a poor miserable wretch, who for many years has been attempting to beat off pain by a constant recurrence to the vice that reproduces it. Conceive a spirit in hell employed in tracing out for others the road to that heaven from which his crimes exclude him ! In short, conceive whatever is most wretched, helpless, and hopeless, and you will form as tolerable a notion of my state as it is possible for a good man to have.

"I used to think the text in St. James, that 'he who offended in one point, offends in all,' very harsh, but I now feel the awful, the tremendous truth of it. In the one crime of OPIUM, what crime have I not made myself guilty of ? Ingratitude to my Maker ! and to my benefactors, injustice ! *and unnatural cruelty to my poor children !*—self-contempt for my repeated promise — breach, nay, too often, actual falsehood.

"After my death, I earnestly entreat that a full and unqualified narration of my wretchedness and of its guilty cause may be made public, that at least some little good may be effected by the direful example.

"May God Almighty bless you, and have mercy on your still affectionate, and in his heart grateful,

"S. T. COLERIDGE.

"JOSIAH WADE, ESQ."

" It appears that in the spring of 1816 Mr. Coleridge left Mr. Morgan's house at Calne, and in a desolate state of mind repaired to London ; when the belief remaining strong on his mind that his opium habits would never be effectually subdued till he had subjected himself to medical restraint, he called on Dr. Adams, an eminent physician, and disclosed to him the whole of his painful circumstances, stating what he conceived to be his only remedy.   The doctor, being a humane man, sympathized with his patient, and knowing a medical gentleman who resided three or four miles from town, who would be likely to undertake the charge, he addressed the following letter to Mr. Gilman :

"'HATTON GARDEN, April 9, 1816.

" ' DEAR SIR :—A very learned, but in one respect an unfortunate gentleman, has applied to me on a singular occasion. He has for several years been in the habit of taking large quantities of opium.   For some time past he has been in vain endeavoring to break himself off it.   It is apprehended his friends are not firm enough, from a dread lest he should suffer by suddenly leaving it off, though he is conscious of the contrary, and has proposed to me to submit himself to any regimen, however severe.   With this view he wishes to fix himself in the house of some medical gentleman, who will have courage to refuse him any laudanum, and under whose assistance, should he be the worse for it, he may be relieved. .As he is desirous of retirement and a garden, I could think of none so readily as yourself.   Be so good as to inform me whether such a proposal is absolutely inconsistent with your family arrangements.   I should not have proposed it, but on account of the great importance of the character as a literary man.   His communicative temper will make his society very interesting as well as useful.   Have the goodness to favor me with an immediate answer; and believe me, dear Sir, your faithful humble servant,

"' JOSEPH ADAMS.' "

Mr. Gilman, in his "Life of Coleridge," says : "I had seen the writer of this letter but twice in my life, and had no intention of receiving an inmate into my house. I however determined on seeing Dr. Adams, for whether the person referred to had taken opium from choice or necessity, to me he was equally an object of commiseration and interest. Dr. Adams informed me that the patient had been warned of the danger of discontinuing opium by several eminent medical men, who at the same time represented the frightful consequences that would most probably ensue. I had heard of the failure of Mr. Wilberforce's case under an eminent physician at Bath, in addition to which the doctor gave me an account of several others within his own knowledge. After some further conversation it was agreed that Dr. Adams should drive Coleridge to Highgate the following evening. On the following evening came Coleridge *himself*, and alone. Coleridge proposed to come the following evening, but he first informed me of the painful opinion which he had received concerning his case, especially from one medical man of celebrity. The tale was sad, and the opinion given unprofessional and cruel, sufficient to have deterred most men so afflicted from making the attempt Coleridge was contemplating, and in which his whole soul was so deeply and so earnestly engaged. My situation was new, and there was something affecting in the thought that one of such amiable manners, and at the same time so highly gifted, should seek comfort and medical aid in our quiet home. Deeply interested, I began to reflect seriously on the duties imposed upon me, and with anxiety to expect the approaching day. It brought me the following letter :

" 'MY DEAR SIR : . . . . And now of myself. My ever-wakeful reason and the keenness of my moral feelings will secure you from all unpleasant circumstances connected with me save only one, viz., the evasion of a specific madness. You will never *hear* any thing but truth from me. Prior habits render it out of my power to tell an untruth, but unless carefully observed, I dare not promise that I should not, with regard to this detested poison, be capable of acting one. No sixty hours have yet passed without my having taken laudanum, though for the last week comparatively trifling doses. I have full belief that your anxiety need not be extended beyond the first week, and for the first week I shall not, I must not, be permitted to leave your house unless with you. Delicately or indelicately, this must be done, and both the servants and the assistant must receive absolute commands from you. The stimulus of conversation suspends the terror that haunts my mind ; but when I am alone the horrors I have suffered from laudanum, the degradation, the blighted utility, almost overwhelm me. If (as I feel for the *first time* a soothing confidence it will prove) I should leave you restored to my moral and bodily health, it is not myself only that will love and honor you ; every friend I have (and, thank God ! in spite of this wretched vice I have many and warm ones, who were friends of my youth and have never deserted me) will thank you with reverence.' "

Dr. Gilman's admiration of Coleridge's talents and respect for his character soon became so enthusiastic that the remainder of the poet's life was made comfortable by his care and under his roof. After the death of Coleridge the first

volume of a biography was published by Dr. G., but has
never been compĺeted.   We are therefore left in ignorance
of the process by which his addiction to opium was reduced
to the small daily allowance which he used during the later
years of his life.   It seems from the following letter addressed to Dr. Gilman more than six years after.he was received
as a member of his household, that the conflict with the
habit was still going on.   " I am still too much under the
cloud of past misgivings—too much of the stun and stupor
from the recent peals and thunder-crash still remain—to
permit me to anticipate others than by wishes and prayers."

Coleridge wrote but little respecting his own infirmity.
Ten years after his domestication in the family of Dr. Gilman he made the following memorandum :

" I wrote a few stanzas twenty years ago—soon after my
eyes had been opened to the true nature of the habit into
which I had been ignorantly deluded by the seeming magic
effects of opium in the sudden removal of a supposed rheumatic affection, attended with swellings in my knees and pal-
pitations of the heart, and pains all over me, by which I had
been bedridden for nearly six months.   Unhappily, among
my neighbor's and landlord's books was a large parcel of
medical reviews and magazines.   I had always a fondness
(a common case, but most mischievous turn with reading men
who are at all dyspeptic) for dabbling in medical writings ; and in one of these reviews I met a case which I fancied very like my own, in which a cure had been affected
by the Kendal Black Drop.   In an evil hour I procured it.
It worked miracles.   The swellings disappeared, the pains
vanished ; I was all alive ; and all around me being as ignorant as myself, nothing could exceed my triumph.   I talked of
nothing else, prescribed the newly-discovered panacea for all
complaints, and carried a bottle about with me, not to lose

any opportunity of administering 'instant relief and speedy cure' to all complainers, stranger or friend, gentle or simple. Need I say that my own apparent convalescence was of no long continuance? But what then? the remedy was at hand and infallible. Alas! it is with a bitter smile, a laugh of gall and bitterness, that I recall this period of unsuspecting delusion, and how I first became aware of the Maelstrom, the fatal whirlpool to which I was drawing just when the current was already beyond my strength to stem. God knows that from that moment I was the victim of pain and terror, nor had I at any time taken the flattering poison as a stimulus, or for any craving after pleasurable sensation. I needed none and oh! with what unutterable sorrow did I read the 'Confessions of an Opium-eater,' in which the writer with morbid vanity makes a boast of what was my misfortune, for he had been faithfully and with an agony of zeal warned of the gulf, and yet willfully struck into the current! Heaven be merciful to him!

"Even under the direful yoke of the necessity of daily poisoning by narcotics, it is somewhat less horrible through the knowledge that it was not from any craving for pleasurable animal excitement, but from pain, delusion, error, of the worst ignorance, medical sciolism, and (alas! too late the plea of error was removed from my eyes) from terror and utter perplexity and infirmity—sinful infirmity, indeed, but yet not a willful sinfulness—that I brought my neck under it. Oh, may the God to whom I look for mercy through Christ, show mercy on the author of the 'Confessions of an Opium-eater,' if, as I have too strong reason to believe, his book has been the occasion of seducing others into this withering vice through wantonness. From this aggravation I have, I humbly trust, been free as far as acts of my free-will and intention are concerned ; even to the author of that work I pleaded with flowing tears, and with an agony of forewarning. He utterly denied it, but I fear that I had even then to *deter*, perhaps not to forewarn."

Referring to the character of Coleridge's disorder, Dr. Gilman says : " He had much bodily suffering. The *cause* of this was the organic change slowly and gradually taking place in the structure of the heart itself. But it was so masked by other sufferings, though at times creating despondency, and was so generally overpowered by the excitement of animated conversation, as to leave its real cause undiscovered." *

In a volume entitled " Letters, Conversations, and Recollections of S. T. C.," written by an intimate friend, we find the following declaration from Coleridge himself :

" My conscience indeed bears me witness, that from the time I quitted Cambridge no human being was more indifferent to the pleasures of the table than myself, or less needed any stimulation to my spirits ; and that, by a most unhappy quackery, after having been almost bedrid for near six months with swollen knees, and other distressing symptoms of disordered digestive functions, and through that most pernicious form of ignorance, medical half-knowledge, I was *seduced* into the use of narcotics, not secretly, but (such was my ignorance) openly and exultingly, as one who had discovered, and was never weary of recommending, a grand panacea, and saw not the truth till my *body* had contracted a habit and a necessity ; and that, even to the latest, my responsibility is for cowardice and defect of fortitude, not for the least craving after gratification or pleasurable sensation of any sort, but for yielding to pain, terror, and haunting bewilderment. But this I say to *man* only, who knows only what has been yielded, not

---

* " *My heart, or some part* about it, seems breaking, as if a weight were suspended from it that stretches it. Such is the *bodily feeling* as far as I can express it by words."—*Coleridge's letter to Morgan.*

H

what has been resisted ; before God I have but one voice—
Mercy ! mercy ! woe is me.

"Pray for me, my dear friend, that I may not pass such
another night as the last. While I am awake and retain my
reasoning powers the pang is gnawing, but I am, except for
a fitful moment or two, tranquil ; it is the howling wilderness
of sleep that I dread." (July 31, 1820.)

From this *bodily* slavery (for it was *bodily*) to a baneful
drug he was never *entirely* free, though the quantity was so
greatly reduced as not materially to affect his health or
spirits.

A good deal that is known respecting Coleridge's opium
habits is derived from the published papers of De Quincey,
whose opportunities for becoming fully informed on the subject are beyond question :

"I now gathered that procrastination in excess was, or
had become, a marked feature in Coleridge's daily life. Nobody who knew him ever thought of depending on any appointment he might make. Spite of his uniformly honorable
intentions, nobody attached any weight to his assurances *in
re futura*. Those who asked him to dinner, or any other party,
as a matter of course sent a carriage for him, and went personally or by proxy to fetch him ; and as to letters, unless
the address was in some female hand that commanded his
affectionate esteem, he tossed them all into one general *dead-
letter bureau*, and rarely, I believe, opened them at all. But
all this, which I heard now for the first time and with much
concern, was fully explained, for already he was under the
full dominion of opium, as he himself revealed to me—
with a deep expression of horror at the hideous bondage—in
a private walk of some length which I took with him about
sunset.

"At night he entered into a spontaneous explanation of

this unhappy overclouding of his life, on occasion of my saying accidentally that a toothache had obliged me to take a
few drops of laudanum. At what time or on what motive he
had commenced the use of opium he did not say, but the
peculiar emphasis of horror with which he warned me against
forming a habit of the same kind, impressed upon my mind
a feeling that he never hoped to liberate himself from the
bondage.

" For some succeeding years he did certainly appear to
me released from that load of despondency which oppressed
him on my first introduction. Grave, indeed, he continued
to be, and at times absorbed in gloom ; nor did I ever see
him in a state of perfectly natural cheerfulness. But as he
strove in vain for many years to wean himself from his cap-
tivity to opium, a healthy state of spirits could not be much
expected. Perhaps, indeed, where the liver and other organs
had for so long a period in life been subject to a continual
morbid stimulation, it may be impossible for the system ever
to recover a natural action. Torpor, I suppose, must result
from continued artificial excitement, and perhaps upon a
scale of corresponding duration. Life, in such a case, may
not offer a field of sufficient extent for unthreading the fatal
links that have been wound about the machinery of health
and have crippled its natural play.

" One or two words on Coleridge as an opium-eater. We
have not often read a sentence falling from a wise man with
astonishment so profound as that particular one in a letter
of Coleridge to Mr. Gilman, which speaks of the effort to
wean one's self from opium as a trivial task. There are, we
believe, several such passages, but we refer to that one in particular which assumes that a single ' week ' will suffice for
the whole process of so mighty a revolution. Is indeed Levia-
than so tamed ? In that case the quarantine of the opium-

eater might be finished within Coleridge's time and with Coleridge's romantic ease. But mark the contradictions of this extraordinary man. He speaks of opium excess, his own excess, we mean—the excess of twenty-five years—as a thing to be laid aside easily and forever within seven days; and yet, on the other hand, he describes it pathetically, sometimes with a frantic pathos, as the scourge, the curse, the one almighty blight which had desolated his life.

"This shocking contradiction we need not press. All will see *that*. But some will ask, was Mr. Coleridge right in either view? Being so atrociously wrong in the first notion (viz., that the opium of twenty-five years was a thing easily to be forsworn), when a child could know that he was wrong, was he even altogether right, secondly, in believing that his own life, root and branch, had been withered by opium? For it will not follow, because, with a relation to happiness and tranquillity, a man may have found opium his curse, that therefore, as a creature of energies and great purposes, he must have been the wreck which he seems to suppose. Opium gives and takes away. It defeats the *steady* habit of exertion, but it creates spasms of irregular exertion; it ruins the natural power of life, but it develops preternatural paroxysms of intermitting power.

"Let us ask any man who holds that not Coleridge himself but the world as interested in Coleridge's usefulness has suffered by his addiction to opium, whether he is aware of the way in which opium affected Coleridge; and secondly, whether he is aware of the actual contributions to literature—how large they were—which Coleridge made *in spite* of opium. All who are intimate with Coleridge must remember the fits of genial animation which were created continually in his manner and in his buoyancy of thought by a recent or an *extra* dose of the omnipotent drug. A lady, who knew nothing experimentally of opium, once told us that she 'could tell when Mr. Coleridge had taken too much opium by his shining countenance.' She was right. We know that mark of opium excesses well, and the cause of it, or at least we believe the cause to lie in the quickening of the insensible perspiration which accumulates and glistens on the face. Be that as it may, a criterion it was that could not deceive us as to the condition of Coleridge. And uniformly in that condition he made his most effective intellectual displays. It is true that he might not be happy under this fiery animation, and we believe that he was not. Nobody is happy under laudanum except for a very short term of years. But in what way did that operate upon his exertions as a writer? We are of opinion that it killed Coleridge as a poet, but proportionably it roused and stung by misery his metaphysical instincts into more spasmodic life. Poetry can flourish only in the atmosphere of happiness, but subtle and perplexed investigation of difficult problems are among the commonest resources for beguiling the sense of misery. It is urged, however, that even on his philosophic speculations opium operated unfavorably in one respect, by often causing him to leave them unfinished. This is true. Whenever Coleridge (being highly charged or saturated with opium) had written with distempered vigor upon any question, there occurred, soon after, a recoil of intense disgust, not from his own paper only but even from the subject. All opium-eaters are tainted with the infirmity of leaving works unfinished and suffering reactions of disgust. But Coleridge taxed himself with that infirmity in verse before he could at all have commenced opium-eating. Besides, it is too much assumed by Coleridge and by his biographer that to leave off opium was of course to regain juvenile health. But all opium-eaters make the mistake of supposing every pain or irritation which they suffer to be the product of opium ; whereas a wise man will say, 'Suppose you do leave off opium, that will not deliver you from the load of years (say sixty-three) which you carry on your back.'

" It is singular, as respects Coleridge, that Mr. Gilman never says one word upon the event of the great Highgate experiment for leaving off laudanum, though Coleridge came to Mr. Gilman for no other purpose ; and in a week this vast creation of new earth, sea, and all that in them is, was to have been accomplished. We *rayther* think, as Bayley junior observes, ' that the explosion must have hung fire.'

" He [Mr. Gilman] has very improperly published some intemperate passages from Coleridge's letters, which ought to have been considered confidential unless Coleridge had left them for publication, charging upon the author of the ' Opium Confessions ' a reckless disregard of the temptations which in that work he was scattering abroad among men. We complain, also, that Coleridge raises a distinction, perfectly perplexing to us, between himself and the author of the ' Opium Confessions ' upon the question—why they severally began the practice of opium-eating. In himself it seems this motive was to relieve pain, whereas the Confessor was surreptitiously seeking for pleasure. Ay, indeed ! where did he learn *that?* We have no copy of the ' Confessions ' here, so we can not quote chapter and verse, but we distinctly remember that toothache is recorded in that book as the particular occasion which first introduced the author to the knowledge of opium. Whether afterward, having been thus initiated by the demon of pain, the opium Confessor did not apply powers thus discovered to purposes of mere pleasure, is a question for himself, and the same question applies with the same cogency to Coleridge. Coleridge began in rheumatic pains. What then ? This is no proof that he did not end in voluptuousness. For our part, we are slow to believe that ever any man did or could learn the somewhat awful truth, that in a certain ruby-colored elixir there lurked a divine power to chase away the genius of ennui, without subsequently abusing this power. True it is that generations have used laudanum as an anodyne (for instance, hospital patients) who have not afterward courted its powers as a voluptuous stimulant ; but that, be sure, has arisen from no abstinence in *them.* There are in fact two classes of temperaments as to this terrific drug—those which are and those which are not preconformed to its power ; those which genially expand to its temptations, and those which frostily exclude them. Not in the energies of the will, but in the qualities of the nervous organization, lies the dread arbitration of—Fall or stand: doomed thou art to yield, or strengthened constitutionally to resist. Most of those who have but a low sense of the spells lying couchant in opium have practically no sense at all ; for the initial fascination is for *these* effectually defeated by the sickness which Nature has associated with the first stages of opium-eating. But to that other class whose nervous sensibilities vibrate to their profoundest depths under the first touch of the angelic poison, opium is the Amreeta cup of beatitude. Now in the original higher sensibility is found some palliation for the *practice* of opium-eating ; in the greater temptation is a greater excuse.

"Originally his sufferings, and the death within him of all hope—the palsy, as it were, of that which is the life of life and the heart within the heart—came from opium. But two things I must add—one to explain Coleridge's case, and the other to bring it within the indulgent allowance of equitable judges. *First*, the sufferings from morbid derangement, originally produced by opium, had very possibly lost that simple character, and had themselves reacted in producing secondary states of disease and irritation, not any longer dependent upon the opium, so as to disappear with its disuse ; hence a more than mortal discouragement to accomplish this disuse when the pains of self-sacrifice were balanced by no gleams of restorative feeling.   Yet, *secondly*, Coleridge did make prodigious efforts to deliver himself from this thraldom ; and he went so far at one time in Bristol, to my knowledge, as to hire a man for the express purpose, and armed with a power of resolutely interposing between himself and the door of any druggist's shop.   It is true that an authority derived only from Coleridge's will could not be valid against Coleridge's own counter-determination : he could resume as easily as he could delegate the power.   But the scheme did not entirely fail.   A man shrinks from exposing to another that infirmity of will which he might else have but a feeble motive for disguising to himself ; and the delegated man, the external conscience as it were of Coleridge, though destined in the final resort, if matters came to absolute rupture—and to an obstinate duel, as it were, between himself and his principal—in that extremity to give way, yet might have long protracted the struggle before coming to that sort of *dignus vindice nodus ;* and, in fact, I know upon absolute proof that before reaching that crisis the man showed fight ; and faithful to his trust, and comprehending the reasons for it, he declared that if he must yield he would ' know the reason why.'

"His inducement to such a step [his visit to Malta] must have been merely a desire to see the most interesting regions of the Mediterranean, under the shelter and advantageous introduction of an official station.   It was, however, an unfortunate chapter of his life ; for being necessarily thrown a good deal upon his own resources in the narrow society of a garrison, he there confirmed and cherished, if he did not there form, his habit of taking opium in large quantities. I am the last person in the world to press conclusions harshly or uncandidly against Coleridge, but I believe it to be notorious that he first began the use of opium not as a relief from any bodily pains or nervous irritations—for his constitution was strong and excellent—but as a source of luxurious sensation. It is a great misfortune, at least it is a great peril, to have tasted the enchanted cup of youthful rapture incident to the poetic temperament. That standard of high-wrought sensibility once made known experimentally, it is rare to see a submission afterward to the sobrieties of daily life. Coleridge, to speak in the words of Cervantes, wanted better bread than was made of wheat; and when youthful blood no longer sustained the riot of his animal spirits, he endeavored to excite them by artificial stimulants.

" Coleridge was at one time living uncomfortably enough at the *Courier* office in the Strand. In such a situation, annoyed by the sound of feet passing his chamber-door continually to the printing-room of this great establishment, and with no gentle ministrations of female hands to sustain his cheerfulness, naturally enough his spirits flagged, and he took more than ordinary doses of opium. Thus unhappily situated, he sank more than ever under the dominion of opium, so that at two o'clock, when he should have been in attendance at the Royal Institute, he was too often unable to rise from bed. His appearance was generally that of a person struggling with pain and overmastering illness. His lips were baked with feverish heat and often black in color, and in spite of the water which he continued drinking through the whole course of his lecture, he often seemed to labor under an almost paralytic inability to raise the upper jaw from the lower.

" But apparently he was not happy himself. The accursed

H 2

drug poisoned all natural pleasure at its sources ; he burrowed continually deeper into scholastic subtleties and metaphysical abstraction ; and, like that class described by Seneca in the luxurious Rome of his days, he lived chiefly by candle-light. At two or three o'clock in the afternoon he would make his first appearance. Through the silence of the night, when all other lights had long disappeared, in the quiet cottage of Grassmere *his* lamp might be seen invariably by the belated traveller as he descended the long steep from Dun-mail-raise, and at five or six o'clock in the morning, when man was going forth to his labor, this insulated son of reveries was retiring to bed."

Those who were nearest and dearest to Coleridge by affection and blood have left on record their sentiments respecting him in the following language. His nephew says : " Coleridge was a student all his life. He was very rarely indeed idle in the common sense of the term, but he was constitutionally indolent, averse from continuous exertion externally directed, and consequently the victim of a procrastinating habit, the occasion of innumerable distresses to himself and of endless solicitude to his friends, and which materially impaired though it could not destroy the operation and influence of his wonderful abilities. Hence also the fits of deep melancholy which from time to time seized his whole soul, during which he seemed an imprisoned man without hope of liberty."

His daughter remarks : " Mr. De Quincey mistook a constitution that had vigor in it for a vigorous constitution. His body was originally full of life, but it was full of death also from the first. There was in him a slow poison which gradually leavened the whole lump, and by which his muscular frame was prematurely slackened and stupefied. Mr. Stuart says that his letters are ' one continued flow of complaint of ill health and incapacity from ill health.' This is true of all his letters (all the *sets* of them) which have come under my eye, even those written before he went to Malta, where his opium habits were confirmed. If my father sought more from opium than the mere absence of pain, I feel assured that it was not luxurious sensations or the glowing phantasmagoria of passive dreams, but that the power of the medicine might keep down the agitations of his nervous system, released for a time at least from the tyranny of ailments which by a spell of wretchedness fix the thoughts upon themselves, perpetually throwing them inward as into a stifling gulf."

Miss Coleridge thus expresses the views of her father's family in respect to Mr. Cottle's publications : " I take this opportunity of expressing my sense of many kind acts and much friendly conduct of Mr. Cottle toward my father, by whom he was ever remembered with respect and affection. If I still regard with any disapproval his publication of letters exposing his friend's unhappy bondage to opium, and consequent embarrassments and deep distress of mind, it is not that I would have wished a broad influencive fact, in the history of one whose peculiar gifts had made him in some degree an object of public interest, to be finally concealed, supposing it to be attested, as this has been, by clear, unambiguous documents. I agree with Mr. Cottle in thinking that he himself would have desired, even to the last, that whatever benefit the world might obtain by the knowledge of his sufferings from opium—the calamity which the unregulated use of this drug had been to him and into which he first fell ignorantly and innocently (not, as Mr. De Quincey has said, to restore the 'riot of his animal spirits' when 'youthful blood no longer sustained it,' but as a relief from bodily pain and nervous irritation) that others might avoid the rack on which so great a part of his happiness for so long

a time was wrecked. Such a wish indeed he once strongly
expressed, but I believe myself to be speaking equally in his
spirit when I say that all such considerations of advantage to
the public should be subordinated to the prior claims of private and natural interests. I should never think the public
good a sufficient apology for publishing the secret history of
any man or woman whatever, who had connections remaining upon earth ; but if I were possessed of private notices
respecting one in whom the world takes an interest, I should
think it right to place them in the hands of his nearest relations, leaving it to them to deal with such documents as a
sense of what is due to the public and what belongs to open-
ness and honesty may demand."

The nephew of Coleridge, in the Preface to the "Table
Talk," says : "A time will come when Coleridge's life may
be written without wounding the feeling or gratifying the
malice of any one ; and then, among other misrepresenta-
tions, that as to the origin of his recourse to opium will be
made manifest ; and the tale of his long and passionate
struggle with and final victory over the habit will form one
of the brightest as well as most interesting traits of the moral
and religious being of this humble, this exalted Christian.

" Coleridge—blessings on his gentle memory !—Coleridge
was a frail mortal. He had indeed his peculiar weaknesses
as well as his unique powers ; sensibilities that an averted
look would rack ; a heart which would have beaten calmly
in the tremblings of an earthquake. He shrank from mere
uneasiness like a child, and bore the preparatory agonies of
his death-attack like a martyr. Sinned against a thousand
times more than sinning, he himself suffered an almost life-
long punishment for his errors, while the world at large has
the unwithering fruits of his labors, his genius, and his sacrifice."

WILLIAM BLAIR.

THE following narrative of a case of confirmed opium-eating was communicated to the editor of the *Knickerbocker Magazine*, in the year 1842, by Dr. B. W. M'Cready of New York, accompanied by the following statement :

POOR BLAIR, whose account of himself I send you, was brought to the City Hospital by a Baptist clergyman in 1835, at which time I was Resident Physician of the establishment. His wretched habit had at that time reduced him to a state of deplorable destitution, and he came to the hospital as much for the sake of a temporary asylum as to endeavor to wean himself from the vice which had brought him to such a condition. When he entered it was with the proviso that he should be allowed a certain quantity of opium per day, the amount of which was slowly but steadily decreased. The dose he commenced with was eighty grains ; and this quantity he would roll into a large bolus, of a size apparently too great for an ordinary person to swallow, and take without any appearance of effort. Until he had swallowed his ordinary stimulus he appeared languid, nervous, and dejected. He at all times had a very pale and unhealthy look, and his spirits were irregular ; although it would be difficult to separate the effects produced by the enormous quantity of opium to which he had been accustomed from the feelings caused in a proud and intellectual man by the utter and irretrievable ruin which he had brought upon himself. Finding him possessed of great information and uncommon ability, I furnished him with books and writing materials, and extended to him many privileges not enjoyed by the ordinary patients in the wards. Observing that he—as is common with most men of a proud disposition who have not met with the success in the world which they deem due to their merits—had paid great attention to his own feelings, I was desirous of having an account written by himself of the effects which opium had produced upon his system. On my making the request he furnished me with the memoir of himself now in your possession. His health at this time was very much impaired. I had been in the habit of giving him orders upon the apothecary for his daily quantum of opium, but when the dose had been reduced to sixteen grains I found that he had counterfeited the little tickets I gave him and thus often obtained treble and quadruple the quantity allowed. After this, of course, although I felt profoundly sorry for the man, the intercourse between us was only that presented by my duty. Shortly afterward he disappeared from the hospital late at night. I have since met him several times in the streets; but for the last three or four years I have neither seen nor heard of him. With his habits it is scarcely probable that he still survives. Poor fellow! He furnishes another melancholy instance of the utter inefficiency of mere learning or intelligence in preserving a man from the most vicious and degrading abuses. He had neither religion nor moral principle; and that kind of gentlemanly feeling which from association he did possess, only made him feel more sensibly the degradation from which it could not preserve him.

BLAIR'S NARRATIVE.

BEFORE I state the result of my experience as an opium-eater, it will perhaps not be uninteresting, and it certainly will conduce to the clearer understanding of such statement, if I give a slight and brief sketch of my habits and history previous to my first indulgence in the infernal drug which has embittered my existence for seven most weary years. The death of my father when I was little more than twelve months old made it necessary that I should receive only such an education as would qualify me to pursue some business in my native town of Birmingham ; and in all probability I should at this moment be entering orders or making out invoices in that great emporium had I not at a very early age evinced an absorbing passion for reading, which the free ac-cess to a tolerably large library enabled me to indulge, until it had grown to be a confirmed habit of mind, which, when the attention of my friends was called to the subject, had be-come too strong to be broken through ; and with the usual foolish family vanity they determined to indulge a taste so early and decidedly developed, in the expectation, I verily believe, of some day catching a reflected beam from the fame and glory which I was to win by my genius ; for by that mystical name was the mere musty talent of a *nelluo librorum* called. The consequence was that I was sent when eight years of age to a public school. I had however before this tormented my elder brother with ceaseless importunity until he had consented to teach me Latin, and by secretly poring over my sister's books I had contrived to gain a tolerable book-knowledge of French.

From that hour my fate was decided. I applied with un-wearied devotion to the study of the classics — the only branch of education attended to in the school—and I even considered it a favor to be allowed to translate, write exercises and themes, and to compose Latin verses for the more idle of my school-fellows. At the same time I devoured all books of whatever description which came in my way—poems, novels, history, metaphysics, or works of science—with an indiscriminating appetite, which has proved very injurious to me through life. I drank as eagerly of the muddy and stagnant pool of literature as of the pure and sparkling fountain glowing in the many-hued sunlight of genius. After two years had been spent in this manner I was removed to another school, the principal of which, although a fair mathematician, was a wretched classical scholar. In fact I frequently construed passages of Virgil, which I had not previously looked at, when he himself was forced to refer to Davidson for assistance. I stayed with him, however, two years, during which time I spent all the money I could get in purchasing Greek and Hebrew books, of which languages I learned the rudiments and obtained considerable knowledge without any instruction. After a year's residence at the house of my brother-in-law, which I passed in studying Italian and Persian, the Bishop of Litchfield's examining chaplain, to whom I had been introduced in terms of the most hyperbolical praise, prevailed on his Diocesan and the Earl of Calthorpe to share the expense of my further education.

In consequence of this unexpected good fortune I was now placed under the care of the Rev. Thomas Fry, rector of the village of Emberton in Buckinghamshire, a clergyman of great piety and profound learning, with whom I remained about fifteen months, pursuing the study of languages with increased ardor. During the whole of that period I never allowed myself more than four hours' sleep ; and still unsatisfied, I very generally spent the whole night, twice a week, in the insane pursuit of those avenues to distinction to which alone my ambition was confined. I took no exercise, and the income allowed me was so small that I could not afford a meat dinner more than once a week, and at the same time set apart the half of that allowance for the purchase of books, which I had determined to do. I smoked incessantly ; for I now required some stimulus, as my health was much injured by my unrelaxing industry. My digestion was greatly impaired, and the constitution of iron which Nature had given me threatened to break down ere long under the effects of the systematic neglect with which I treated its repeated warnings. I suffered from constant headache ; my total inactivity caused the digestive organs to become torpid ; and the unnutritious nature of the food which I allowed myself would not supply me with the strength which my assiduous labor required. My nerves were dreadfully shaken, and at the age of fourteen I exhibited the external symptoms of old age. I was feeble and emaciated ; and had this mode of life continued twelve months longer, I must have sank under it.

I had during these fifteen months thought and read much on the subject of revealed religion, and had devoted a considerable portion of my time to an examination of the evidences advanced by the advocates of Christianity, which resulted in a reluctant conviction of their utter weakness and inability. No sooner was I aware that so complete a change of opinion had taken place, than I wrote to my patron, stating the fact and explaining the process by which I had arrived at such a conclusion. The reply I received was a peremptory order to return to my mother's house immediately ; and on arriving there, the first time I had entered it for some years, I was met by the information that I had nothing more to expect from the countenance of those who

had supplied me with the means of prosecuting my studies
to "so bad a purpose." I was so irritated by what I considered the unjustifiable harshness of this decision, that at the
moment I wrote a haughty and angry letter to one of the
parties, which of course widened the breach and made the
separation between us eternal.

What was I now to do? I was unfit for any business, both
by habit, inclination, and constitution. My health was ruined, and hopeless poverty stared me in the face ; when a distinguished solicitor in my native town, who by the way has
since become celebrated in the political world, offered to receive me as a clerk. I at once accepted the offer; but
knowing that in my then condition it was impossible for me
to perform the duties required of me, I decided on TAKING
OPIUM ! The strange confessions of De Quincey had long
been a favorite with me. The first part of it had in fact
been given me both as a model in English composition and
also as an exercise to be rendered into Patavinian Latin. The
latter part, the " Miseries of Opium," I had most unaccountably always neglected to read. Again and again, when my
increasing debility had threatened to bring my studies to an
abrupt conclusion, I had meditated this experiment, but an
undefinable and shadowy fear had as often stayed my hand.
But now that I knew that unless I could by artificial stimuli
obtain a sudden increase of strength I must STARVE, I no
longer hesitated. I was desperate ; I believed that something
horrible would result from it ; though my imagination, most
vivid, could not conjure up visions of horror half so terrific
as the fearful reality. I knew that for every hour of comparative ease and comfort its treacherous alliance might confer
upon me *now*, I must endure days of bodily suffering ; but I
did not, could not conceive the mental hell into whose fierce,
corroding fires I was about to plunge.

All that occurred during the first day is imperishably engraved upon my memory. It was about a week previous to the day appointed for my debut in my new character as an attorney's clerk; and when I arose, I was depressed in mind, and a racking pain to which I had lately been subject, was maddening me. I could scarcely manage to crawl into the breakfast-room. I had previously procured a drachm of opium, and I took two grains with my coffee. It did not produce any change in my feelings. I took two more—still without effect; and by six o'clock in the evening I had taken ten grains. While I was sitting at tea I felt a strange sensation, totally unlike any thing I had ever felt before ; a gradual *creeping thrill*, which in a few minutes occupied every part of my body, lulling to sleep the before-mentioned racking pain, producing a pleasing glow from head to foot, and inducing a sensation of dreamy exhilaration (if the phrase be intelligible to others as it is to me), similar in nature but not in degree to the drowsiness caused by wine, though not inclining me to sleep ; in fact so far from it that I longed to engage in some active exercise—to sing or leap. I then resolved to go to the theatre—the last place I should the day before have dreamed of visiting ; for the sight of cheerfulness in others made me doubly gloomy. I went, and so vividly did I feel my vitality—for in this state of delicious exhilaration even mere excitement seemed absolute Elysium—that I could not resist the temptation to break out in the strangest vagaries, until my companions thought me deranged. As I ran up the stairs I rushed after and flung back every one who was above me. I escaped numberless beatings solely through the interference of my friends. After I had become seated a few minutes, the nature of the excitement was changed, and a " waking sleep" succeeded. The actors on the stage vanished ; the stage itself lost its ideality ; and be-

fore my entranced sight magnificent halls stretched out in
endless succession, with gallery above gallery, while the roof
was blazing with gems like stars whose rays alone illumined
the whole building, which was thronged with strange, gigantic
figures—like the wild possessors of a lost globe, such as Lord
Byron has described in "Cain" as beheld by the fratricide,
when, guided by Lucifer, he wandered among the shadowy ex-
istences of those worlds which had been destroyed to make
way for our pigmy earth.   I will not attempt further to describe the magnificent vision which a little pill of "brown
gum" had conjured up from the realm of ideal being.   No
words that I can command would do justice to its Titanian
splendor and immensity.

At midnight I was roused from my dreamy abstraction;
and on my return home the blood in my veins seemed to
"run lightning," and I knocked down (for I had the strength
of a giant at that moment) the first watchman I met.   Of
course there was a row, and for some minutes a battle-royal
raged in New Street, the principal thoroughfare of the town,
between my party and the " Charlies," who, although greatly
superior in numbers, were sadly " milled," for we were all
somewhat scientific bruisers—that sublime art or science having been cultivated with great assiduity at the public school
through which I had, as was customary, fought my way.   I
reached home at two in the morning with a pair of " Oxford
spectacles " which confined me to the house for a week.   I
slept disturbedly, haunted by terrific dreams, and oppressed
by the nightmare and her nine-fold, and awoke with a dreadful headache ; stiff in every joint, and with deadly sickness of
the stomach which lasted for two or three days ; my throat
contracted and parched, my tongue furred, my eyes bloodshot, and the whole surface of my body burning hot.   I did
not have recourse to opium again for three days ; for the

strength it had excited did not till then fail me. When partially recovered from the nausea the first dose had caused, my spirits were good, though not exuberant, but I could eat nothing and was annoyed by an insatiable thirst. I went to the office, and for six months performed the services required of me without lassitude or depression of spirits, though never again did I experience the same delicious sensations as on that memorable night which is an "oasis in the desert" of my subsequent existence; life I can not call it, for the "*vivida vis animi et corporis*" was extinct.

In the seventh month my misery commenced. Burning heat, attended with constant thirst, then began to torment me from morning till night; my skin became scurfy; the skin of my feet and hands peeled off; my tongue was always furred; a feeling of contraction in the bowels was continual; my eyes were strained and discolored, and I had unceasing headache. But internal and external heat was the pervading feeling and appearance. My digestion became still weaker, and my incessant costiveness was painful in the extreme. The reader must not however imagine that all these symptoms appeared suddenly and at once; they came on gradually, though with frightful rapidity, until I became a "*morborum moles*," as a Roman physician whose lucubrations I met with and perused with great amusement some years since in a little country ale-house poetically expresses it. I could not sleep for hours after I had lain down, and consequently was unable to rise in time to attend the office in the morning, though as yet no visions of horror haunted my slumbers. Mr. P., my employer, bore with this for some months; but at length his patience was wearied, and I was informed that I must attend at nine in the morning. I could not; for even if I rose at seven, after two or three hours unhealthy and fitful sleep, I was unable to walk or exert myself in any way for at least two hours. I was at this time taking laudanum, and had no appetite for any thing but coffee and acid fruits. I could and did drink great quantities of ale, though it would not, as nothing would, quench my thirst.

Matters continued in this state for fifteen months, during which time the only comfortable hours I spent were in the evening, when freed from the duties of the office I sat down to study, which it is rather singular I was able to do with as strong zest and as unwearied application as ever; as will appear when I mention that in those fifteen months I read through in the evenings the whole of Cicero, Tacitus, the Corpus Pœtarum (Latinorum), Boëthius, Scriptores Historiæ Augustinæ, Homer, Corpus Græcarum Tragediarum, a great part of Plato, and a large mass of philological works. In fact, in the evening I generally felt comparatively well, not being troubled with many of the above-mentioned symptoms. These evenings were the very happiest of my life. I had ample means for the purchase of books, for I lived very cheap on bread, ale, and coffee, and I had access to a library containing all the Latin classics—Valpy's edition in one hundred and fifty volumes, octavo, a magnificent publication—and about fifteen thousand other books. Toward the end of the year 1829 I established at my own expense, and edited myself, a magazine (there was not one in a town as large and as populous as New York!) by which I lost a considerable sum ; though the pleasure I derived from my monthly labors amply compensated me. In December of that year my previous sufferings became light in comparison with those which now seized upon me, never completely to leave me again. One night, after taking about fifty grains of opium, I sat down in my arm-chair to read the confession of a Russian who had murdered his brother because he was the chosen of her whom both loved. It was recorded by a French priest who visited

him in his last moments, and was powerfully and eloquently
written.   I dozed while reading it; and immediately I was
present in the prison-cell of the fratricide.   I saw his ghastly
and death-dewed features; his despairing yet defying look;
the gloomy and impenetrable dungeon; the dying lamp, which
seemed but to render darkness visible; and the horror-struck
yet pitying expression of the priest's countenance; but there
I lost my identity.   Though I was the recipient of these impressions, yet I was not myself separately and distinctively
existent and sentient; but my entity was confounded with
that of not only the two figures before me, but of the inani-
mate objects surrounding them.   This state of compound
existence I can no further describe.   While in this state I
composed the " Fratricide's Death," or rather it composed itself and forced itself upon my memory without any activity
or volition on my part.

And here again another phenomenon presented itself.
The images reflected (if the expression be allowable) in the
verses rose bodily and with perfect distinctness before me,
simultaneously with their verbal representations; and when I
roused myself (I had not been *sleeping*, but was only *abstracted*)
all remained clear and distinct in my memory.   From that
night, for six months, darkness always brought the most horrible fancies, and opticular and auricular or acoustical delu-
sions of a frightful nature, so vivid and real that instead of
a blessing, sleep became a curse, and the hours of darkness
became hours which seemed days of misery.   For many consecutive nights I dared not undress myself nor put out the
light, lest the moment I lay down some "*monstrum horren-
dum, informe, ingens*" should blast my sight with his hellish
aspect!   I had a double sense of sight and sound; one real,
the other visionary; both equally strong and apparently real;
so that while I distinctly heard imaginary footsteps ascending

the stairs, the door opening and my curtains drawn, I at the same time as plainly heard any actual sound in or outside the house, and could not remark the slightest difference between them ; and while I *saw* an imaginary assassin standing by my bed, bending over me with a lamp in one hand and a dagger in the other, I could see any real tangible object which the degree of light which might be then in the room made visible.   Though these visionary fears and imaginary objects had presented themselves to me every night for months, yet I never could convince myself of their non-existence ; and every fresh appearance caused suffering of as intense and as deadly horror as on the first night !   So great was the confusion of the real with the unreal that I nearly became a convert to Bishop Berkeley's non-reality doctrines. My health was also rapidly becoming worse ; and before I had taken my opium in the morning I had become unable to move hand or foot, and of course could not rise from my bed until I had received strength from the " damnable dirt." I could not attend the office at all in the morning, and was forced to throw up my articles, and, as the only chance left me of gaining a livelihood, turn to writing for magazines for support.

I left B. and proceeded to London, where I engaged with Charles Knight to supply the chapters on the use of elephants in the wars of the ancients for the " History of Elephants," then preparing for publication in the series of the Library of Entertaining Knowledge.   For this purpose I obtained permission to use the library of the British Museum for six months, and again devoted myself with renewed ardor to my favorite studies.

" But what a falling off was there ! "   My memory was impaired, and in reading I was conscious of a confusion of mind which prevented my clearly comprehending the full meaning of what I read. Some organ appeared to be defective. My judgment too was weakened, and I was frequently guilty of the most absurd actions, which at the time I considered wise and prudent. The strong common sense which I had at one time boasted of, deserted me. I lived in a dreamy, imaginative state which completely disqualified me for managing my own affairs. I spent large sums of money in a day, and then starved for a month ; and all this while the " *chateux en Espagne,*" which once only afforded me an idle amusement, now usurped the place of the realities of life and led me into many errors, and even unjustifiable acts of immorality, which lowered me in the estimation of my acquaintances and friends, who saw the effect but never dreamed the cause. Even those who knew I was an opium-eater, not being aware of the effect which the habitual use of it produced, attributed my mad conduct to either want of principle or aberration of intellect, and I thus lost several of my best friends and temporarily alienated many others. After a month or two passed in this employment I regained a portion of strength sufficient to enable me to obtain a livelihood by reporting, on my own account, in the courts of law in Westminster, any cause which I judged of importance enough to afford a reasonable chance of selling again; and by supplying reviews and occasional original articles to the periodicals, the *Monthly,* the *New Monthly, Metropolitan,* etc. My health continued to improve, probably in consequence of my indulging in higher living and taking much more exercise than I had done for two or three years ; as I had no need of buying books, having the use of at least five hundred thousand volumes in the Museum. I was at last fortunate enough to obtain the office of Parliamentary reporter to a morning paper, which produced about three hundred pounds a year ; but after working on an average fourteen or fifteen hours

I

a day for a few months, I was obliged to resign the situation
and again depend for support on the irregular employment I
had before been engaged in, and for which I was now alone fit.
My constitution now appeared to have completely sunk under
the destroying influence of the immense quantity of opium I
had for some months taken—two hundred, two hundred and
fifty, and three hundred grains a day. I was frequently
obliged to repeat the dose several times a day, as my stomach
had become so weak that the opium would not remain upon
it ; and I was besides afflicted with continual vomiting after
having eaten any thing. I really believed that I could not last
much longer. Tic-douloureux was also added to my other
suffering ; constant headache, occasional spasms, heart-burn,
pains in the legs and back, and a general irritability of the
nerves, which would not allow me to remain above a few
minutes in the same position. My temper became soured
and morose. I was careless of every thing, and drank to
excess in the hope of thus supplying the place of the stimulus which had lost its power.

At length I was compelled to keep my bed by a violent
attack of pleurisy, which has since seized me about the same
time every year. My digestion was so thoroughly ruined
that I was frequently almost maddened by the sufferings
which indigestion occasioned. I could not sleep, though I
was no longer troubled with visions, which had left me about
three months. At last I became so ill that I was forced to
leave London and visit my mother in Kenilworth, where I
stayed ; writing occasionally, and instructing a few pupils
in Greek and Hebrew. I was also now compelled to sell
my library, which contained several Arabic and Persian
MSS., a complete collection of Latin authors, nearly a complete one of Greek, and a large collection of Hebrew and
Rabbinic works, which I had obtained at a great expense and

with great trouble. All went. The only relics of it I was
able to retain were the "Corpus Pœtarum, Græcarum et
Latinorum," and I have never since been able to collect another library. Idleness, good living, and constant exercise
revived me ; but with returning strength my nocturnal visit-
ors returned, and again my nights were made dreadful. I
was terrified through visions similar to those which had so
alarmed me at first, and I was obliged to drink deeply at
night to enable me to sleep at all. In this state I continued
till June, 1833, when I determined once more to return to
London, and I left Kenilworth without informing any one of
my intention the night before. The curate of the parish
called at my lodging to inform me that he had obtained the
gift of six hundred pounds to enable me to reside at Oxford
until I could graduate. Had I stayed twenty-four hours
longer I should not now be living in hopeless poverty in a
foreign country ; but pursuing, under more favorable auspices
than ever brightened my path before, those studies which
supported and cheered me in poverty and illness, and with a
fair prospect of obtaining that learned fame for which I had
longed so ardently from my boyhood, and in the vain endeav-
or to obtain which I had sacrificed my health and denied myself not only the pleasures and luxuries but even the necessaries of life. I had while at the office in B. entered my
name on the books of Brazen-nose College, Oxford, and resided there one term, not being able to afford the expense attendant on a longer residence. Thus it has been with me
through life. Fortune has again and again thrown the
means of success in my way, but they have always been like
the waters of Tantalus—alluring but to escape from my grasp
the moment I approached to seize them.

I remained in London only a few days, and then proceeded to Amsterdam, where I stayed a week, and then went to

Paris. After completely exhausting my stock of money I was compelled to walk back to Calais, which I did with little inconvenience, as I found that money was unnecessary ; the only difficulty I met with being how to escape from the over-flowing hospitality I everywhere experienced from rich and poor. My health was much improved when I arrived in town, and I immediately proceeded on foot to Birmingham, where I engaged with Dr. Palmer, a celebrated physician, to supply the Greek and Latin synonyms and correct the press for a dictionary of the terms used by the French in medicine, which he was preparing. The pay I received was so very small that I was again reduced to the poorest and most meagre diet, and an attack of pleurisy produced such a state of debility that I was compelled to leave Birmingham and return to my mother's house in Kenilworth.

I had now firmly resolved to free myself from my fatal habit ; and the very day I reached home I began to diminish the quantity I was then taking by one grain per day. I received the most careful attention, and every thing was done that could add to my comfort and alleviate the sufferings I must inevitably undergo. Until I had arrived at seventeen and a half grains a day I experienced but little uneasiness, and my digestive organs acquired or regained strength very rapidly. All constipation had vanished. My skin became moist and more healthy, and my spirits instead of being depressed became equable and cheerful. No visions haunted my sleep. I could not sleep, however, more than two or three hours at a time, and from about 3 A.M. until 8—when I took my opium—I was restless and troubled with a gnawing, twitching sensation in the stomach. From seventeen grains downward my torment (for by that word alone can I characterize the pangs I endured) commenced. I could not rest, either lying, sitting, or standing.

I was compelled to change my position every moment, and the only thing that relieved me was walking about the country. My sight became weak and dim; the gnawing at my stomach was perpetual, resembling the sensation caused by ravenous hunger; but food. though I ate voraciously, would not relieve me. I also felt a sinking in the stomach, and such a pain in the back that I could not straighten myself up. A dull, constant, aching pain took possession of the calves of my legs, and there was a continual jerking motion of the nerves from head to foot. My head ached, my intellect was terribly weakened and confused, and I could not think, talk, read, nor write. To sleep was impossible, until by walking from morning till night I had so thoroughly tired myself that pain could not keep me awake, although I was so weak that walking was misery to me. And yet under all these *désagrèmens* I did not feel dejected in spirit; although I became unable to walk, and used to lie on the floor and roll about in agony for hours together. I should certainly have taken opium again if the chemist had not, by my mother's instructions, refused to sell it. I became worse every day, and it was not till I had entirely left off the drug—two months nearly—that any alleviation of my suffering was perceptible. I gradually but very slowly recovered my strength both of mind and body, though it was long before I could read or write, or even converse. My appetite was too good; for though while an opium-eater I could not endure to taste the smallest morsel of fat, I now could eat at dinner a pound of bacon which had not a hair's-breadth of lean in it. Previously to my arrival in Kenilworth an intimate friend of mine had been ruined—reduced at once from affluence to utter penury by the villainy of his partner, to whom he had entrusted the whole of his business, and who had committed two forgeries for which he was sentenced to

transportation for life. In consequence of this event, my
friend, who was a little older than myself and had been
about twelve months married, determined to leave his young
wife and child and seek to rebuild his broken fortunes in
Canada. When he informed me that such was his plan I
resolved to accompany him, and immediately commenced
preparations for my voyage. I was not however ready,
not having been able so soon to collect the sum necessary, when he was obliged to leave, and as I could not
have him for my companion, I altered my course and took
my passage for New York, in the vain expectation of obtaining a better income here, where the ground was comparatively unoccupied, than in London, where there were
hundreds of men as well qualified as myself, dependent on
literature for their support. I need not add how lamenta-
bly I was disappointed. The first inquiries I made were
met by advice to endeavor to obtain a livelihood by some
other profession than authorship. I could get no employment as a reporter, and the applications I addressed to
the editors of several of the daily newspapers received no
answer. My prospects appeared as gloomy as they could
well be, and my spirits sunk beneath the pressure of the
anxious cares which now weighed so heavily upon me. I
was alone in a strange country, without an acquaintance
into whose ear I might pour the gathering bitterness of my
blighted hopes. I was also much distressed by the intense
heat of July, which kept me from morning till night in a state
much like that occasioned by a vapor bath. I was so melancholy and hopeless that I really found it necessary to have
recourse to brandy or opium. I preferred the latter, although to ascertain the difference, merely as a philosophical experiment, I took rather copious draughts of the former
also. But observe ; I did not intend ever again to become

the slave of opium. I merely proposed to take three or four grains a day until I should procure some literary engagement, and until the weather became more cool. All my efforts to obtain such engagement were in vain ; and I should undoubtedly have sunk into hopeless despondency had not a gentleman (to whom I had brought an order for a small sum of money, twice the amount of which he had insisted on my taking), perceiving how injuriously I was affected by my repeated disappointments, offered me two hundred dollars to write " Passages from the Life of an Opium-eater," in two volumes. I gladly accepted this disinterested offer, but before I had written more than two or three sheets I became disgusted with the subject. I attempted to proceed, but found that my former facility in composition had deserted me ; that, in fact, I could not write. I now discovered that the attempt to leave off opium again would be one of doubtful result. I had increased my quantum to forty grains. I again became careless and inert, and I believe that the short time that had elapsed since I had broken the habit in England had not been sufficient to allow my system to free itself from the poison which had been so long undermining its powers. I could not at once leave it off; and in truth I was not very anxious to do so, as it enabled me to forget the difficulties of the situation in which I had placed myself; while I knew that with regained freedom the cares and troubles which had caused me again to flee to my destroyer for relief, would press upon my mind with redoubled weight. I remained in Brooklyn until November. Since then, I have resided in the city, in great poverty, frequently unable to procure a dinner, as the few dollars I received from time to time scarcely sufficed to supply me with opium. Whether I shall now be able to leave off opium, God only knows !

THE manuscript of the narrative which follows was placed in the hands of the compiler by a physician of Philadelphia who for many years had shown great kindness to its writer, in the endeavor to cure him of his pernicious habits. The writer seems from childhood to have been cursed with an excessive sensibility, and an unusual constitutional craving for excitement, coupled with an infirm and unreliable will. The habit of daily dependence upon alcohol appears to have been established for years before the use of opium was commenced ; and the latter was begun chiefly for the purpose of substituting the excitement of the drug in place of the excitement furnished by brandy and wine. That any human being can permanently substitute the daily use of the one in place of the daily use of the other is more than doubtful. Attempts of this kind are not unfrequently made, but the result is uniformly the same—a double tyranny is established which no amount of resolution is sufficient to conquer. This fact is so forcibly illustrated in this autobiography, that although it is chiefly a story of suffering from the use of alcoholic stimulants, its insertion here may serve as a caution to that class of persons, not inconsiderable in number, who are tempted to substitute one ruinous habit in place of another.

I am inclined to think I must have been born, if not literally with a propensity to *stimulus*, at least with a susceptibility to fall readily into the use of it ; for my ancestors, so far as I know, all used alcohol, though none of them, I believe, died drunkards. One of my earliest recollections is that of seeing the tumbler of sling occasionally partaken of by the elders of the family, even before breakfast, and of myself with the other children being sometimes gratified with a spoonful of the beverage or the sugar at the bottom. Paregoric, too—combining two of the most dangerous of all substances, alcohol and opium—was a favorite medicine of my excellent mother, and in all the little ailments of childhood was freely administered. So highly thought she of it that on my leaving home at fifteen for Cambridge University she put a large vial of it in my trunk, with the injunction to take of it, if ever sick.

In my young days I saw alcohol used everywhere. How in those days any body failed of the drunkard's grave seems hardly less than miraculous. How I myself escaped becoming inebriate for more than twenty-five years, is with my organization, a deep mystery.

I can remember, when quite young, occasionally drinking —as I saw every body else do, boys as well as men, and even women—and I recollect also being two or three times overcome with liquor, to my infinite horror and shame not less than bodily suffering. At fifteen, as I said, I entered Harvard University, perfectly free from the *habit* of drinking as from all other bad habits. Here too, as everywhere before, I saw alcohol flowing copiously, the most prevalent kind being wine.

On Exhibition and Commencement Days, every student honored with a " part " was accustomed at his room to make his friends and acquaintances free of the cake-basket and especially of the wine-cup. A good deal of wine and punch too was drank at the private " Blows " (so called) of the stu-

I 2

dents, at the meetings of their various clubs, at their military musterings, and other like occasions. At all such times there was more or less intoxication. I can remember being a good deal disordered with wine two or three times during my four college years, and I have no doubt I was considerably affected by it more times than these; still scholastic ambition, somewhat diligent habits of study, straitened means, and the want of any special inclination for artificial stimulus carried me through college without my having contracted any habit of drinking or having grown to depend at all upon stimulants.

But deteriorating causes had been at work, and though the volcano had not burst forth as yet, the material had been silently gathering through these four seemingly peaceful years. In the winter of my sixteenth or seventeenth year, after suffering several days from severe toothache, I was induced by my landlady, a pipe-smoker, to try tobacco as a remedy. The result of this trial, which proved effectual, was that partly from the old notion that tobacco was a teeth-preservative, and partly, I suppose, because the taste was hereditary, I fell at once into the habit of tobacco-chewing, which I continued without intermission for eleven years. In this abominable practice I exercised no moderation: indeed in any practice of this kind it has seemed constitutional with me to go to excess, and unnatural to pursue a middle course. None at all or too much was the alternative exacted by my organization. By consequence, the perpetual, unmeasured waste of saliva induced by using such immoderate quantities of this weed must speedily have exhausted a constitution not endowed with unusual vital energies. As it was I must have received deep injury. I often felt faintness and languor, though I did not or would not admit what now I have no doubt of—that this vegetable was in fault.

At nineteen, graduating at Cambridge, I took and kept for the three following years an academy in a near neighboring town.   Here I soon began to suffer (what I now suppose) the ill effects of the false education and false living (the to-bacco-chewing, physical inertness, mental partialness, and the rest) of long foregoing years.   I began to suffer greatly from gloom and depression of spirits.   Short fits of morbid gayety and long stretches of dullness and darkness made up the present, while the future looked almost wholly black.   I had indeed been afflicted so long as I could remember with seasons of low spirits, but *these* glooms, for depth and long continuance, transcended any thing I had ever experienced before. On festive occasions, at which I was often present, I was accustomed to take a glass or half-glass of wine with and like the rest ; but other than this, I used no stimulus and never had thought of keeping any at my lodgings.   In fact, so little was I *seasoned* in this way that half a glass of ordinary wine was enough to elevate my spirits many degrees above their usual pitch.   I know not why it never occurred to me to use habitually what I found occasionally to be such a relief. A few months after commencing school I attended with a party of friends the celebration of the Landing of the Pilgrims at Plymouth.   The orator was exceedingly eloquent ; the occasion one of great enthusiasm ; and what with my intense previous excitement of mind, what with my unseasoned brain, and what with the universal example of the wise and good about me, I took so much wine at the public dinner as to be completely intoxicated, and was only able after three or four hours of sleep to attend the Pilgrim Ball.   My shame, remorse, and horror on this occasion was so far salutary that without any special resolution I was for a long time after, a total abstinent.   In fact this monitory influence lasted with more or less force for six or seven years.   But the gloom and depression before spoken of came to a crisis. About a year after my leaving college I broke down with a severe attack of dyspepsia. A weight pressing continually on my chest, palpitation of the heart, sleeplessness by night, or dreams that robbed sleep of all repose, debility, languor, and increased gloom—such are some of the symptoms that hung oppressively upon me for more than a year.

Under these circumstances I took a physician's advice. By his orders I swallowed I know not how many bottles of bitters. Whether from their effect or from Nature's curative power in despite of them, my ailments at last mostly disappeared ; but to this very hour I have been more or less subject to the same physical inertness and unexcitability, low spirits, and many like symptoms. No unexperienced person can imagine what a life it is to be thus physically but half alive. The temptation is incessant to raise by artificial helps the physical tone, in order thus to attain activity and energy of mind. My only wonder is that I did not sooner resort to what would at least give temporary relief to the depression and torpor from which I suffered so much and so long.

After keeping school three years, being the last of the three a member of the Cambridge Divinity School, I passed two years at that school and was licensed to preach. My life there was the same false, unnatural one it had been in college—much study and no bodily exercise, a few faculties active and the greater number exercised scarce at all. All this while, with the exception of tobacco, I used no stimulants except on rare occasions, and then always in moderation.

In August, 1829, I was licensed as a preacher by the Boston Ministerial Association. In the December following I was ordained a minister at Lynn, Mass. In May, 1830, I was married, and in the succeeding autumn became a housekeeper. Immediately on becoming an ordained clergyman I procured one or two demijohns of wine as a preparative for hospitality to my clerical brethren and to visitants generally. Such was the custom universally, and in various ways I was given to understand that I too must adopt it. Keeping wine at home now for the first time, I tasted it doubtless oftener than ever before, though still not habitually or with any approach to excess. Furthermore, a member of my family, in debilitated health and a dyspeptic, was ordered by the family physician, one of the most distinguished of the Boston Faculty, to take brandy and water with dinner as a tonic. A demijohn of brandy therefore took its place in the closet beside the demijohn of wine already there, and on the daily dinner-table was set a decanter of this liquid fire. For myself I had as already intimated never perfectly recovered from my ancient dyspeptic attack, nor was my present way of life very favorable to health. To replenish this waste, a good deal of bodily exercise was needed, but of such exercise I took scarce any at all.

It was then no uncommon thing for a minister to sit down on Saturday evenings with a pot of green tea as strong as lye, or of coffee black as ink, and a box of cigars beside him —drinking at the one and puffing at the other all or most of the night through—and under the excitement of these nerve-rasping substances trace rapidly on paper the words which next day were to thrill or melt his listeners. A final cup of tea or coffee, extra strong, and a last cigar before entering the pulpit, gave him that fervor and unction of manner so indispensable to eloquence. His theme, perhaps, was intemperance ; and with nerves tingling from the action of liquids which no swine will drink, and of the plant which no swine will eat, he would portray most vividly the terrible ruin wrought by intoxicating drink. Do not believe, however, that in all this he was dishonest or hypocritical ; he was merely self-ignorant—blind to the fact that in condemning the alcoholic inebriate he was by every word condemning himself as well.　This ignorance, however, could not obviate the effects of such hideous outrage on the physical laws.

I have dwelt on these points partly for their intrinsic truth and importance, and partly as bearing upon and explaining my own case.　In ill health, languid and restless from the causes pertaining to my then condition, I found in brandy or wine a temporary relief for that languor and sedative for that restlessness.　When necessitated to write, and the mind was dull because the body was sluggish, instead of seeking the needed life in tea and coffee and tobacco-smoking, I found it more readily in brandy or wine.　In short, I began somewhat to depend on these stimulants for the excitement I required for my work.　I hardly need say I dreamed of neither wrong nor danger in so doing, and it was yet a good while before a case of intoxication awoke me from this false security.　Thus three years passed, at the close of which I removed to Brookline for the health of a friend apparently declining in consumption.　Just before leaving I cast away the tobacco which I had used largely for ten or eleven years.　The struggle was a hard one, and the faintness and uneasy cravings which long tormented me operated, I think, as a temptation to replace the lost stimulus by increased quantities of alcoholic stimulus.

Under these circumstances I went to Brookline in the beginning of February, 1833, and for three or four months I shut myself up as sole attendant and nurse of a sick friend, apparently dying.　I had no external employment compelling my attention ; there were no outward objects to call me off from my infirmities and uneasy sensations.　I was alone with all these—alone with sickness and coming death—alone with a gloomy present and a clouded future—and the bottle stood near, promising relief. It is not very strange that I resorted oftener than before to its treacherous comfort, and became more than ever accustomed to depend upon it. I believe, however, that only once during these months was I positively overcome by it, and I was very ready to cheat myself into the belief that other causes were in fault besides, and as much as alcohol. The ensuing summer I spent partly in Cambridge and partly in travelling with the invalid who still survived ; and with health considerably improved I continued stimulus, though I think in rather less quantities than in the winter preceding. Once, however, I was badly intoxicated with port wine, and so ill as greatly to alarm my friends and induce them to call in a physician, who administered a powerful emetic. Whether or not he understood the nature of my ailment I never knew. My friends I think did not, and I was very willing to cheat myself into the belief that the wine thus affected me because I was ill from other causes.

At the close of August of this year I went to Brooklyn, New York, to preach for a few Sundays to a handful of persons who had just united to attempt forming a new religious society. I remained through the winter following. A society was gathered ; I was installed over it, and there continued till the summer of 1837. These four years were to me tremendous years. They seem to me, in looking back, like a long, sick, feverish dream. Even now I can hardly but shudder at the remembrance of glooms of midnight blackness and sufferings that mock all endeavors at description : for it absolutely appears to me on the review that not for one week of these four years was I a free, healthful, sober man ; not one week but I was rent by a fierce conflict between "the law of the members and the law of the mind." How it was I executed the amount I did, of intellectual labor—how it was I accomplished the results I did, is to me an impenetrable mystery. I began to address in a hired school-house a handful of persons, having most of them but a slight mutual acquaintance, and in my farewell discourse I addressed a fair-sized, closely-united congregation assembled in their own conveniently-spacious church, with the organization and all the customary belongings of the oldest worshiping societies. Not one Sunday of that time was I disenabled by my fatal habits to perform the customary offices ; but I did not understand my condition in any thing like its reality as now I look back upon it. My actual state was known to but very few in its entireness—I may say to absolutely none of those I daily companied with—and I did at the close of that period receive an honorable dismissal at my own request, a request made for reasons distinct from this ; nor between myself and people, or any of them, was there ever a word exchanged on this subject from first to last. "Truth is strange, stranger than fiction."

I shall not attempt going through these years in detail. I went to Brooklyn with the habit of depending on alcohol to a considerable extent for physical tone and mental excitement, though not with the *habit* of losing my balance thereby.

It was some time after establishing myself in New York before I became at all awake to my condition. At considerable intervals I had two or three attacks of convulsionary fits. My physician gave them some name—I hardly remember what—but he did not specify the cause. I now understand them to have been intoxication fits. I suspected then that alcohol had some connection with them, and I was so far aroused to this and other evils of my way of life that I attempted total abstinence. But besides a host of uneasy sensations, I at once experienced such a lack of bodily strength and of mental life and activity that to think or write, or apply myself to my tasks generally, I found impossible.

After making several abortive attempts of this kind, I tried at last the substitution of laudanum for alcohol. It was a most fatal move! for the final result was a bondage of which previously I had not even a conception. At first, however, I seemed as though lifted out of the pit into Paradise. Instead of the feverish, tumultuous excitement of alcohol, I experienced a calm, equable, thrilling enjoyment. My whole being was exalted from its previous turmoil and perturbation and heat, to dwell in a region of serenity and peace and quiet bliss. But alas for the reverse side of the picture! The total prostration, the depth of depression, the more than infantile feebleness following the reaction of this excitement—the multitude of uneasy, uncomfortable, often bewildering sensations pertaining to the habit, are such as can not be conveyed to one inexperienced in the matter. But any one may decide that the presence and incorporation with the system, in large quantities, of a poison which is so deadly a foe to life and all life's movements can not be without very marked and baneful results. The fact is that there is not one out of the thousand various functions of the body which is not deranged and turned away by this cause, and the movements of the mind and heart are from sympathy hardly less morbid. Whether such a state must not be one of sufferings many, and often frightful, every one may judge.

But worse even than this followed. It was not very long before the opium nearly lost its power to excite and enliven, though it still kept an inexorable clutch on every fibre of my frame, and I was compelled to take it daily to keep the very current of life flowing.

To make my condition worse still, while obliged to use opium daily to prolong even this existence—gloomy and

apathetic as it was—I found that in order to think or work
with any thing of vigor I absolutely required, every now and
then, some excitement which opium now would not give.   I
tried, therefore, strong tea and coffee and tobacco-smoking.
But all these were not enough, and I found there was nothing for me but to try alcohol again ; so that the upshot of
my experiment of substituting opium for alcohol was, that I
got opium, alcohol, tea, coffee, and tobacco-smoking fastened upon me all at once and all in excessive quantities ; and
the consequence of using alcohol was that no caution I could
employ would secure me from occasional intoxication.   Such
was my physical derangement that I never could be certain
beforehand of the degree of effect which alcoholic stimulus
would exert upon me, and the same quantity which at one
time would produce only the excitement I sought, would under other physical conditions completely overcome me.

During my last two years in Brooklyn I made several attempts to break away from opium and other stimulus, and
each time made considerable progress.   But the same circumstances yet existed that originally led to the evil, and
in fact others of the same class had been superadded, while
the whole operated with aggravated force, so that I found or
thought it impossible to achieve my freedom without dis-
closing my state, and thus, as I supposed, setting the seal to
my own temporal ruin.   Once and again, therefore, I went
back to my dungeon.

It may here be remarked that the sedentary man has extraordinary difficulties to contend with in such a case.   His
occupation being lonely, and demanding no bodily exertion,
he has little or nothing to draw off, *perforce*, his attention
from the innumerable aches and tormenting sensations which
beset him, sometimes for months without cessation, in going
through the extricating process.   To sit still and endure long-

protracted torment demands a resolution compared with which the courage that carries one into a battle-field is a paltry thing.

But this bondage so galling, this position so false in all ways, and so severely condemned alike by conscience and honor, determined me at last to attempt my freedom at the cost even of life, if need be. I broke up housekeeping, sent my family away, and commenced the struggle. I had a bad cold at the time, besides a complication of various cares and distresses which probably increased the severity of the trial. Violent brain-fever came on, accompanied with universal inflammation and a host of sensations for which I never could find any name. It seemed as if my arteries and veins ran with boiling water instead of blood, and as the current circulated through the brain I felt as if it actually boiled up against and tossed the skull at the top of my head, as you have seen the water in a tea-kettle rattling the lid. My hearing was affected in a thousand strange ways : I heard a swimming noise which went monotonously on for weeks without cessation. The ocean, with all its varieties of sound, was forever in my hearing. Sometimes I heard the long billowy swell of the sea after a hard blow ; again I could hear the sharp, fuming collision of waves in a storm ; and then for hours I would listen to the solemn, continuous roar, intermitted with the booming, splashing wash of the tempest-roused surge upon the beach. Almost incessantly, too, I heard whisperings, sharp and hissing, on every side—outside and inside of my room—and the whisperers I imagined were all saying hard things of myself.

Meantime my mind was under tremendous excitement, and all its faculties, especially the imagination, were preternaturally active, vivid, and rapid-working. Such was my mental excitement and bodily irritation that for ten days and nights I slept hardly at all, nor enjoyed one moment's release from pain. That I was thoroughly in earnest in what I had undertaken will appear from the fact that all this time I had in a drawer within reach a bottle of laudanum, which I knew would in a few moments give me ease and sleep. Yet thus agonized and half delirious, I notwithstanding left it untouched.

I was mostly confined to the house about four weeks. The inflammation gradually subsiding left me as weak as a child—so morbidly sensitive that tears flowed on the slightest occasion, and with my whole frame pervaded by a dull, incessant ache. To these symptoms were added coldness of the extremities, an obstinate determination of blood to the head, which swelled the vessels of the face and brain almost to bursting, susceptibility to fatigue on the least exertion, physical or mental, and so great a confusion and wandering of thought that it was only by a violent effort that my mind could be brought to act continuously or with the least vigor.

As soon as I was able to go abroad I joined my family in the neighborhood of Boston, in the hope of benefiting by change of scene. Remaining here for several months without much improvement of health, I felt called on for various reasons to resign my charge in New York. Thus left with a family and very slender resources, I was compelled, feeble as I was, to bestir myself for their and my own support. No employment offered itself but that of my profession, and unfit, therefore, as I felt myself, body and mind, for this, I saw no alternative but to preach as occasion presented. It was a most cruel necessity, for without some artificial aid I was unable even to stand through the pulpit services. As a choice of evils I used wine and brandy; for the terrors of opium were still too recent.

In the closing part of December, 1837, I went to the city of Washington to preach for six or seven Sundays. The same necessity, real or supposed, of stimulating, followed me through the six weeks of my stay there. One day at the close of this period, feeling unusually ill and languid, I sent a servant out for a bottle of brandy. I remember pouring out and drinking a single glass of it, and this is the last and whole of my recollection for two days. I awoke and was told I had been exceedingly ill. I must have been very badly intoxicated, though how or why I was so, I know not to this day. So soon as I could hold up my head I went by invitation to Baltimore, and stayed there some three weeks with a college friend. While there I learned from various sources that I was at last palpably and generally exposed and disgraced. I relinquished my profession at once both in reality and name, deeming this the least I could do in the circumstances. About the middle of March, 1838, with shattered, miserable health, overwhelmed with regret and shame and remorse, and the future palled with funereal black, I set out for the residence of relatives in Vermont. Here I remained two and a quarter years, studying law with my sister's husband, who was an attorney and counsellor. For several months I used no stimulus except tobacco, which in the desperate restlessness of the previous summer I had again began to chew after four years' interruption. I of course was weak and languid from this great abstraction of stimulus, coupled with the effects of the severe illness I had undergone. This debility rendered more severe the endurance of other evils of my condition. No wonder that under such wear and tear my nervous system should have become shattered. I was attacked with tic-douloureux. Though suffering severely, old recollections gave me such dread of anodyne and tonic medicines—which I thought it most likely would be administered —that I delayed for some time seeking medical advice. Pain, however, at last drove me to it, and from two physicians I received a prescription of morphine and quinine. I knew that morphine was a preparation of opium, but supposing it a preparation leaving out the stimulating and retaining only the sedative properties of the drug, I imagined it less dangerous than crude opium. With this opinion—with excruciating pain on one side and on the other relief in the physicians' prescription—it is not very strange I chose relief. I used the morphine until apparently the neuralgic affection was cured. On attempting then to lay it aside I found the habit of stimulating again fastened upon me. Once more I found myself neither more nor less than a bond slave to opium to all intents and purposes. With my existing physical debility, with a pressing host of perplexities and tribulations, and with my appalling remembrances of the former struggle, I could not summon resolution and perseverance enough to achieve a second emancipation. So regulating the quantity as well as I could, I waited in hope of some more auspicious season for the attempt.

In the latter part of June, 1840, I went to New York city to complete my third year of legal study. I was at the time weak in body and low-spirited, and my debility was increased by the extraordinary heat of the weather. I was disappointed too in several arrangements on which I had reckoned. The result of all this was a want of physical and moral energy which precluded the attempt at emancipation from opium which I had purposed to make on my arrival; and worse than this, I found myself rapidly getting into the way of adding brandy to opium to procure the desired amount of excitement, as had formerly been the case. I came to the conclusion that I could not achieve my freedom alone, but must have help. I had no home, and after casting about I could devise no better scheme than to enter the Insane Hospital at Bloomingdale. I accordingly went there and stayed thirteen weeks. I found on arriving, that neither myself nor the friends I had advised with had understood the conditions of a residence in that Institution ; for to their disappointment and mine I was locked into the lunatic ward and at first even had a maniac for a room-mate. Being, from the total abandonment of stimulus, in a state of intense nervous excitement, I was for several days, especially during nights, kept on the very verge of frenzy by the mutterings and gibberings, the howlings and horrid execrations of the mad creatures, my neighbors. Without occupation for mind or body—with all things disturbing about me—with deeply depressing remembrances, and the future showing black as midnight—I remained here three months, and it is marvellous that these causes alone did not utterly destroy me. But to fill up the measure, I was attacked with fever and ague, which kept me burning and freezing, shaking and aching, for several weeks, and reduced me to such a degree of feebleness that I kept my bed most of the time. Thus I left the Institution more shattered physically than when I entered—so shattered that it was full two years before I regained my customary measure of bodily strength.

It being now the first of December, 1840, I entered a law office in Wall Street, where I remained till the following July. For some months I enjoyed a glimpse of sunshine and had the hope of being established in business by my employer. But in the spring of 1841 his business fell off so largely that he dismissed three clerks who were there on my entering, and counselled me to seek some more promising sphere. Thus I was again afloat, knowing not whither to turn, and so discouraged as to care little what became of me. One thing only seemed stable and permanent, and that was the temptation to seek a temporary exhilaration in my depression, and a brief oblivion of my troubles, in alcohol.

By another change, in the fore part of July, 1841, I entered Judge Allen's office in Worcester, Mass., and continuing there until March, 1842, was formally admitted to the Bar and commissioned as Justice of the Peace for Essex County. My life in Worcester was pretty regular, though I was not perfectly abstinent, nor did I escape being once or twice overcome.   In March, 1842, I went to Lynn, Mass., as editor of the *Essex County Washingtonian.*   Here was the spot where, technically speaking, I had first entered life, and it was teeming with a thousand memories, now most painful and sad.   Much as I had known before of mental suffering, I can remember none more intense than I experienced the first few months of my return to Lynn.   At times I felt as if any thing were preferable to what I endured, and that to procure relief by any means whatever was perfectly justifiable, on the ground of that necessity which is above all laws. I therefore used morphine, first occasionally and at last habitually, and sometimes, though rarely, brandy.   Some six months after settling in Lynn, being one day in Boston on business, I was oppressed with deadly nausea, for which after trying two or three glasses of plain soda-water as a remedy, I tried a glass of brandy with the soda.   I was made intoxicated by the means and badly so.   I was perplexed as to what I ought to do under the circumstances, but by the advice of two Washingtonians, one of them the general agent of my paper, I still continued at my post of editor.

In the following winter I was up as one of three candidates for Congress from Essex County.   In addition to the usual butting a candidate gets on such occasions—being the third, whose votes prevented a choice of either the other two candidates—I was exposed to a raking fire from the two great political parties.   Out of old truths twisted and exaggerated out of all identity, and new lies coined for the occasion, a world of falsity as to my character and habits was bandied about ; and although a caucus sitting in examination two long successive evenings pronounced the charges against me slanderous and wicked, and published a hand-bill to that effect, yet the proprietor of my paper, moved by a power behind the throne, chose that my connection with the paper should terminate. For some time previous, I had been getting interested in the Association doctrines of Fourier. I now became one of the editors of a monthly magazine devoted in part to the advocacy of these doctrines, which after issuing three numbers was compelled to stop for want of support. I then in September, 1843, went forth on a tour through Massachusetts to lecture on the subject. I thus spent five months, visiting twenty towns and delivering some ninety gratuitous lectures. During this time I used morphine habitually, and occasionally, though rarely, took brandy. I took enough, however, of the latter to partly intoxicate me three or four times, and sufficiently often to prevent the reputation of being intemperate from ever dying away.

Sick and tired out with an existence so false and wretched, I determined again to achieve emancipation at whatever cost, and by the help of Providence, and the kind co-operation of inestimable friends, I succeeded. I suffered severely, but far less than might have been supposed. Cold water, under God, was the great instrument of my cure. Drinking copiously of it, and lying some hours per day swathed in a sheet dipped in it, for about one month, I found the painful symptoms mostly gone ; and three or four months of rest completed the restoration of my strength.

And thus, after years of pain and sufferings in every kind, and errors many and great, I find myself, by God's blessing, free and healthy, and with a youthful life and feeling of which the very memory was almost extinct.

K.

Within a few months from the time this autobiography closes, the writer again relapsed into the use of opium, and was received as a patient into the New York Hospital. While there he furnished the editor of the *Medical Times*, then on duty at the Hospital, with a brief history of his case, substantially agreeing with what has already been given. A portion of the paper is occupied with a comparison of the effects of opium and alcohol on the system, and is valuable as being the experience of one who was eminently familiar with both :

The difference between opium and alcohol in their effects on body and mind, is (judging from my own experience) very great. Alcohol, pushed to a certain extent, overthrows the balance of the faculties, and brings out some one or more into undue prominence and activity ; and (sad indeed) these are most commonly our inferior and perhaps lowest faculties. A man who, sober, is a demi-god, is, when drunk, below even a beast. With opium (*me judice*) it is the reverse. Opium takes a man's mind where it finds it, and lifts it *en masse* on to a far higher platform of existence, the faculties all retaining their former relative positions—that is, taking the mind as it is, it intensifies and exalts all its capacities of thought and susceptibilities of emotion. Not even this, however, extravagant as it may sound, conveys the whole truth. Opium weakens or utterly paralyzes the lower propensities, while it invigorates and elevates the superior faculties, both intellectual and affectional. The opium-eater is without sexual appetite ; anger, envy, malice, and the entire hell-brood claiming kin to these, seem dead within him, or at least asleep ; while gentleness, kindness, benevolence, together with a sort of sentimental religionism, constitute his habitual frame of mind. If a man has a poetical gift, opium almost irresistibly

stirs it into utterance   If his vocation be to write, it matters
not how profound, how difficult, how knotty the theme to be
handled, opium imparts a before unknown power of dealing
with such a theme ; and after completing his task a man
reads his own composition with utter amazement at its
depth, its grasp, its beauty, and force of expression, and won-
ders whence came the thoughts that stand on the page before
him.   If called to speak in public, opium gives him a copiousness of thought, a fluency of utterance, a fruitfulness of
illustration, and a penetrating, thrilling eloquence, which oft-
en astounds and overmasters himself, not less than it kin-
dles, melts, and sways the audience he addresses.   I might
dilate largely on this topic, but space and strength are alike
lacking.

Taking up his personal story where his " Autobiography"
leaves it, and where, as he imagined, hydropathic treatment
had effected a cure, the writer explains how he became for
the third time an opium-eater :

The time came at last when I must work, be the consequences what they would, and work, too, with my brain, my
only implement ; and that time found my brain impotent from
a yet uninvigorated nervous system.   If I would work, I must
stimulate; and morphine, bad as it was, was better than alcohol.   I took morphine once more, and lectured on literary
topics for some months with triumphant success.   While so
lecturing in a country town, I was solicited to take a parish in
the neighborhood.   I did so, and there continued two years
and a quarter, performing in that time as much literary labor
as ever in three times the interval in any prior period of my
life.   In short, I had three happy, intellectually-vigorous, out-
pouring years, with bodily health uniformly sound and com-

plete with the exceptions hereafter to be mentioned. And yet, through those years I never used less than a quarter of an ounce of morphine per week, and sometimes more. I attribute my retaining so much health, in spite of the morphine, to the rigorous salubrity of my habits, bodily and mental, in other respects. Once, and often twice a day, the year round, I laved the whole person in cold water with soap ; I slept with open window the year through excepting stormy winter nights ; I laid upon a hard bed, guiltless of feathers ; I used a simple diet ; and finally, I cherished all gentle and kindly, while rigidly excluding from my mind all bitter and perturbing, feelings. But not to dilate further on mere narrative, let me say that I have continued to use opium, for the most part habitually, from my last assumption of it up to the period of my admission into this Hospital. A year since, however, I dropped morphine, and have since used the opium pill in its stead, sometimes taking an ounce per week, but generally not overpassing a half ounce per week.

And here I may make the general remark, proved true from my own experience, that for all the desirable effects sought from this species of stimulus, a half ounce of gum opium is about the same as an ounce or any larger quantity of said gum, and nearly the same as a quarter-ounce of morphine or more—that is, half an ounce of opium stimulates and braces me at least nearly if not entirely as much as I can be stimulated and braced by this drug. All that is taken over this tends rather to clog, to stupefy, to nauseate, than to stimulate.

Another point in my own experience is, that in a few weeks only, after commencing or recommencing the use of opium, I always reached the full amount which, as a habit, I ever used—that is, either a half-ounce of opium or a quarter-ounce of morphine. I never went on increasing the dose in order to get the required amount of stimulation, but at one or the other of these two points I would remain for years successively. A third remark I would make is, that it is only for the first few weeks after commencing the use of opium that one feels palpably and distinctly the thrilling of the nerves, the sensation of being stimulated and raised above the previously existing physical tone, for which the drug was first taken. All the effects produced after that by the opium, are to keep the body at that level of sensation in which one feels positively alive and capable to act, without being impeded or weighed down by physical languor and impotence. Such languor and impotence one feels from abstaining merely a few hours beyond the wonted time of taking the dose. It is not pleasure, then, that drives onward the confirmed opium-eater, but a necessity scarce less resistible than that Fate to which the pagan mythology subjected gods not less than men.

Let me now, before closing, attempt briefly to describe the effects of opium upon the body and mind of the user, as also the principal sensations accompanying the breaking of the habit.

The opium-eater is prevailingly disinclined to, and in some sort incapacitated for, bodily exertion or locomotion. A considerable part of the time he feels something like a sense, not very distinctly defined, of bodily fatigue ; and to sit continuously in a rocking or an easy chair, or to recline on a sofa or bed, is his preference above all modes of disposing of himself. To walk up a flight of stairs often palpably tires the legs, and makes him pant almost as much as a well person does after pretty rapid motion. His lungs manifestly are somehow obstructed, and do not play with perfect freedom. His liver too is torpid, or else but partially active ; for if using laudanum or the opium pill, he is constantly more or less costive, the fæces being hard and painful to expel; and if using morphine, though he may have a daily movement, yet the fæces are dry and harder than in health. One other morbid physical symptom I remember to have experienced for a considerable time while using a quarter of an ounce of morphine per week, and this was an annoying palpitation of the heart. I was once told, too, by a keen observer, who knew my habit, that my color was apt to change frequently from red to pale.

These are substantially all the physical peculiarities I experienced during my opium-using years. It is still true, however, that the years of my using opium (or, in perfect strictness, morphine) were as healthy as any, if not the very healthiest, of the years of my life.

But what of the effects of opium-eating on the mind? The one great injury it works, is (I think) to the will, that force whereby a man executes the work he was sent here to do, and breasts and overcomes the obstacles and difficulties he is appointed to encounter, and bears himself unflinchingly amid the tempests of calamity and sorrow which pertain to the mortal lot. Hardihood, manliness, resolution, enterprise, ambition, whatever the original degree of these qualities, become grievously debilitated if not wholly extinct. Reverie, the perusal of poetry and fiction, becomes the darling occupation of the opium-user, and he hates every call that summons him from it. Give him an intellectual task to accomplish; place him in a position where a mental effort is to be made; and, most probably, he will acquit him with unusual brilliancy and power, supposing his native ability to be good. But he can not or will not seek and find for himself such work and such position. He feels helpless, and incompetent to stir about and hold himself upright amid the jostling, competitive throngs that crowd the world's paths, and there seek life's prizes by performing life's duties and executing its requisitions. Solitude, with his books, his dreams and imaginings, and the excited sensibilities that lead to no external action, constitute his chosen world and favorite life. In one word, he is a species of maniac; since, I believe, his views, his feelings, and his desires in relation to most things are peculiar, eccentric, and unlike those of other men, or of himself in a state of soundness. There is, however, as complete a "method in his madness" as in the sanity of other men. He is in a different sphere from other men, and in that sphere he is sane.

The first symptoms attendant on breaking off the habit, coming on some hours after omitting the wonted dose, are a constant propensity to yawn, gape, and stretch, together with somewhat of languor, and a general uneasiness. Time passes, and there follows a sensation as if the stomach was drawn together or compressed, as if with a slight degree of cramp, coupled with a total extinction of appetite; the mouth and throat become dry and irritated; there is an incessant disposition to clear the throat by "hemming" and swallowing, and there is a tickling in the nose which necessitates frequent sneezing, sometimes a dozen or even twenty times in succession. As the hours go on, shudders run through the frame, with alternate fever heats and icy chills, hot sweats and cold clammy sweats, while a dull, incessant ache pervades the bones, especially at the joints, alternated by an occasional sharp, intolerable pang, like tic-douloureux. Then follow a host of indescribable sensations, as of burning, tinglings, and twitchings, seeming to run along just beneath the surface of the skin over the whole body, and so strange are these sensations that one is prompted to scream, and strike the wall, the bed, or himself, to vary them. By this time the liver commences a most energetic action, and a violent diarrhea sets in. The discharges are not watery or mucous, but, save in thinness, not very unlike healthy stools for the most part. Not long, however, after the commencement of the diarrhea, so copious is the effusion of bile from the liver, that one will sometimes pass, for a dozen stools in succession, what seems to be merely a blackish bile, without a particle of fæces mingled with it. But this lasts not many days, and is followed by the thin, not altogether unhealthy-looking discharges above mentioned, repeated often an incredible number of times per day. Whether from the quality of these discharges, or from whatever cause, the interior surface of the bowels feels intolerably hot, as though excoriated, and it seems as if boiling water or aqua fortis running through the intestines would scarce torture one more than these stools. In fact, all the internal surfaces of the body are in this same burning, raw-feeling state. The brain, too, is in a highly excited, irritable condition ; the head sometimes aching and throbbing, as though it must burst into fragments, and a humming, washing, simmering noise going on incessantly for days together. Of course there can be no sleep, and one will go on for ten days and nights consecutively without one moment's loss of intensest consciousness, so far as he can judge ! Strange to say, notwithstanding this excessive irritation of the entire system, one feels so feeble and strengthless that he can scarce drag one foot after the other, and to walk a few rods, or up a flight of stairs, is so terribly fatiguing that one must needs sit down and pant. (Let it be noted, that these symptoms belong to the case where one is simply deprived at once and wholly of opium without any medical help, unless the use of cold water be considered such.) These symptoms (unaided by medicine) last, with gradual abatements of virulence, from twenty to thirty days, and then mostly die away. Not well and right, however, does one feel, even then. Though for the most part free from pain, he is yet physically weak, and all corporeal exertion is a distressing effort. He must needs sleep, too, enormously, going to bed often at sunset in a July day, and sleeping log-like until six or seven next morning, and then sleeping with like soundness two or three hours after dinner. How long it would be before the recovery of his complete original strength and natural physical tone, personal experience does not enable me to say. His condition, both in itself and as relates to others, is meanwhile most strange and anomalous. He looks, probably, better than ever in his life before. In sufficiently full flesh, with ruddy cheeks and skin clear as a healthy child's, the beholder would pronounce him in the height of health and vigor, and would glow with indignation at seeing him loitering about day after day, doing little save sleep, in a world where so much work needs to be done. And yet he feels all but impotent for enterprise, or any active physical efforts ; for there is scarce enough nervous force in him to move his frame to a lingering walk, and sometimes it seems as if the nervous fibres were actually pulled out, and he must move, if at all, by pure force of volition.

Most singular too, the while, is the state of his mind. His power of thought is keen, bright, and fertile beyond example, and his imagination swarms with pictures of beauty, while his sensitiveness to impressions and emotions of every kind is so excessively keen that the tears spring to his eyes on the slightest occasion. He is a child in sensibility, while a youth in the vividness, and a man in the grasp, the piercingness and the copiousness of his thoughts. He can not write down his thoughts, for his arm and hand are unnerved ; but in conversation or before an audience he can utter himself as if filled with the breath of inspiration itself.

K 2

INSANITY AND SUICIDE FROM AN ATTEMPT TO ABAN-
DON MORPHINE.

THE account which follows is abridged from advance proof-sheets of a narrative, written for separate publication, by Dr. L. Barnes, of Delaware, Ohio, by whose courtesy a portion of his article appears in these pages.

In the afternoon of Saturday, January 25th, 1868, Rev. G. W. Brush, of Delaware, a clergyman of estimable character and more than respectable talents, was found to have committed suicide. Sixteen or seventeen years previous to this fatal act, morphine had been prescribed to Mr. Brush for occasional disorder in the bowels and for a dormant cancer of the tongue. But something else which had not been prescribed—an unrelenting necessity to go on as he had begun—was also developed in his nature, which in time bore its matured and inevitable fruit. Mr. Brush made his case known for the first time to Dr. Barnes in November, 1866, when his habitual consumption of morphine varied from twelve to fifteen grains daily, with an occasional use of double this quantity.

At this time, in the language of Dr. Barnes, he appeared greatly depressed, mourned over his life as a failure, and said he had been tempted to end it. He had once made a serious effort to abandon the habit, but the effect was so prostrating, and diarrhea, pouring like a flood, had borne him so near the gates of death, that he was compelled to resume the drug in order to save his life. But he was determined to make another attempt, and wished my professional services against the consequences which he well knew must follow.

He entered upon the trial, reducing rapidly the amount of his morphine. I called on him in the course of two or three days, according to appointment, and found him wan and hag-, gard, weak and almost wild with suffering. His hands, lips, and voice trembled. He tottered on his legs ; and, though sweating profusely, he hovered about the fire to keep warm. Day followed day, while he still suffered and endured. On one occasion, as I entered, he had been writing, and read me his production. It was an account of the effects produced by morphine, the giving way of nerves, softening of the muscles, the depression, nightmare in the day-time, visions, horrid shapes ; how the victim is sometimes engulfed in a flood of waters, while faces in all imaginary varieties of distortion, grin from the waves, and terrible eyes gleam forth from their depths.

About this time, business which he thought could not be transacted in his suffering condition unexpectedly demanded his attention, and the attempt was abandoned.

The year 1867 passed with him amid depression, shame, and remorse. He called on me perhaps a hundred times at my office, and seldom left without referring in some way to what he considered his degradation. He repeatedly inquired if I thought it of any use for him to try going on any longer in his ministerial work. Once he came with a brighter face than usual, saying he had concluded to try it one year more, and if he could not succeed—— Then what? I inquired as he paused. A dark cloud spreading over his brow was his only answer, and he lapsed into despondency. This despondency appears to be the legitimate effect of opium. This fact was strikingly manifest in the case of Mr. Brush, for his nåtural disposition, from childhood up, had been usually kind, cheerful, and good ; nor had he any dyspeptic or bilious tendencies to worry and sour him. Few men have ever been physically so well organized, or socially and religiously so well situated for the enjoyment of a prosperous and happy life.

He came to me, finally, on the first day of January, 1868, saying his people had kindly granted him leave of absence for a few weeks, which he would devote to the work of overcoming his enemy, if such a thing were possible. He could not live in his bondage. His wretched life, with its terrible end, was forever staring him in the face. He asked me if I would receive him at my house, and take care of him during the struggle, as I had once consented to do. I said I would if he would consent to let the people know why he was there. He looked very sad as he answered that it would not do. He must undertake the battle at home. He then took from his pocket some papers of morphine, which he had caused to be weighed in doses diminishing at the rate of half a grain each, beginning with six grains for the first day, five and a half for the next, and so on, down. This was a sudden falling off of nearly two-thirds from his ordinary allowance. He gave me all but the two largest powders, which he reserved for an absence of two days at Columbus. He proposed going away for the purpose of coming home sick, in which condition he well knew he should be at that time. I was to call at his house on the evening of his return, to render such assistance as his condition might demand.

I went at the time appointed and found him again shattered, trembling, sweating, and hovering about the fire. He said he had slept none, was suffering much, and that his knees especially were aching badly. He called pleadingly for the amount of morphine prepared for that day, as he had not taken it. It was given, and then he conversed freely for an hour or so.

The next evening he proposed to reduce his morphine by two grains instead of half a grain, but was in a hurry for the quantity he was to have. In the course of over two days more he came down to about two grains for the whole day. But one evening, when I found him apparently much relieved from suffering, and he saw my look of wonder and doubt, he confessed having broken over the rules by taking an additional dose of about three grains on his own responsibility. He said his diarrhea had returned, the medicine left to check it was gone, he hated to send for me, and so had done it. He was full of remorse, declaring that if I should now abandon him, he would not blame me. I told him I should stick to him as long as he would let me ; that he was doing a great work, such as few men ever succeeded in—a work for two worlds, this one and the next—and that he must not give it up.

I continued to spend the evenings with him for about two weeks. The morphine was reduced to something like one grain a day, his appetite returned, and he began to sleep pretty well at night. His nerves became steady, and his diarrhea was controlled without serious difficulty. Energy and strength returned so rapidly that in about two weeks he was ready to resume his work. He said to his wife that the awful weight was all gone—all gone. He expressed his gratitude to me in the most glowing terms. He was triumphant at the idea of having conquered with so much less suffering than he expected. Alas ! I knew his danger, and saw with sorrow that his returning confidence was removing him from under my control while yet the enemy remained in the field.

His last visit to me was on Friday, January 17th. He wanted diarrhea medicine enough to last till the next Tuesday, when he would call again and report. I felt uneasy about him, and went to hear him preach on the intervening Sunday evening. I saw by his flushed and embarrassed manner that he was falling back, and have since learned that after service he confessed to his wife, who was watching his condition with keen eyes, that he had taken about three grains to strengthen him for the occasion. Poor man! He doubtless thought he could stop there. Tuesday came, but he came not to my office. Wednesday, and he came not. Then I was called away from home and did not return until late Saturday night. The first news which greeted me on arriving was, that he was no more. He had been buying morphine at the drug store during the week, and had reached nearly his former quantity. He had wandered about, uncertain, forlorn, desolate. On Friday he had tried to borrow a gun to shoot rats, had come across the way to my office, which was found closed, and then tried again to borrow the gun. He told his wife that dreadful load had come back. Saturday his Quarterly Meeting commenced. He was to preach in the afternoon. He was exceedingly kind and helpful to his family at dinner-time, as he had been all day. The people were assembling at the church, not far off. He went to the barn, suspended a rope from a beam overhead, as he stood upon the manger. It was not quite long enough. He lengthened it with his pocket-handkerchief, looped it around his neck, put his hands in his pockets, and leaped off.

He was gone forever. He had failed in his last attempt to break away from the benumbing power of opium, and in his desperation had sought freedom in death. Let no man judge him, and least of all those who are strangers to the fascinating and infernal strength of his enemy. You may call it a grave mistake, a dreadful blunder, a doleful insanity, but do not assume to put him beyond the reach of mercy,

or to decide that his lamentable end was not the iron door
through which he may have passed to the city of the golden
streets.

A newspaper account of the death of Mr. Brush having
fallen under the notice of a morphine sufferer in Wisconsin,
the latter addressed a letter to Dr. Barnes, in which he gives
his own remarkable experience in the immediate and absolute abandonment of the habit.

The writer is represented as being about fifty years of age,
temperate in his general habits, and though not possessed of great vigor of constitution, as having been through life
a hard-working man.　His use of morphine began in the
year 1861, under a medical prescription for the relief of general debility; but without any knowledge on his part of the
character of the remedy he was using.　After six months
habituation, the attempt to relinquish it proved a failure.
For the first two years, morphine appeared to benefit him.
At the expiration of this time his daily allowance had become three grains, which quantity was rarely exceeded during the four subsequent years of his bondage.　After narrating the mental and physical suffering he underwent in these
years, he says:

April 17, 1867, found me a poor, wasted, miserable, six
years' morphine-eater; health all gone; unable to do any sort
of business; desiring nothing but death to close my sufferings.　Then I made up my mind to stop the use of morphine
all at once.　I had previously attempted to break off by degrees, but I was beaten at that game every time.　It is utterly impossible to taper off by less and less, unless some
one is over the patient watching every motion.　I say it un-
derstandingly—the will of no man is strong enough to handle

the poison for himself. He will make a virtue out of necessity, and for this time will over-take.

So I resolved to quit at once and forever. I arranged my business as far as I could, under the idea that I should die in the attempt. The first forty-eight hours I slept most of the time, waking somewhat often, however, and then dropping asleep, while a sort of nervous twitching would come and go. But the next day found me wide awake. And—shall I tell you?—there was no more sleep for me until sixty-five days had passed. No, not one single moment for sixty-five days and nights. I was fully awake—never slept one moment! The second day my suffering was intense. Every nerve seemed to be on a rampage. Every faculty, mental and physical, appeared to be striving to see how much suffering I could stand. The third day my bowels began to empty, and a river of old fœtid matter ran away. It seemed that I was passing off in corruption. This continued for nearly four long, suffering weeks. I never checked it, but let Nature take her course.

During the first four weeks of the fight there was extreme pain in every part of my body. It seemed to me that I should burn up. This worse than death sensation never left me a single hour for the first thirty-five days. It seemed at times as though my bones would burst open : a sort of nerve fire seemed to be shut up in them which must be let out. I was able to walk out, and if necessary could walk a mile or more.

The fifty-sixth day of suffering without sleep found me at a Water Cure. Warm baths, sometimes with battery, then packs, then sitz baths, for ten more long, suffering days and nights—but sleep never came to me and pain never left me. On the sixty-fifth day of the fight I felt perfectly easy. All my pains were gone. I went to my room and slept nearly four hours. For ten minutes after waking I never stirred a limb or muscle, fearing it would bring back the pains. But a happier man never woke from sleep. I saw that I was delivered from the prison-house of death. I telegraphed to my family that sleep had come. To my dying-hour I shall ever remember that eventful day. But it was only the glimmering of light. Gradually and slowly sleep came to be my companion again. And even yet it has not fully come. Until within the last twenty days when I awoke, every nerve, every emotion was awake all at once.

It is now the tenth month since I quit morphine. Then my weight was only one hundred and twenty-five pounds. Now it is one hundred and ninety. I am the happiest man on the earth, I am redeemed from one of the lowest hells in all worlds.

In a subsequent letter to Dr. Barnes the writer says: " My health still improves. There is one peculiarity about my will-power ; it is so vacillating, not reliable and firm as before. Still I feel that it will come back."

The following declaration, which Dr. Barnes embodies in his article, is deserving the careful consideration both of physicians and philanthropists. He says : " Calling to mind what has come to my knowledge during a long and extensive medical practice, the conclusion is, that I have known of more deaths from the use of opium, in some of its forms, than from all the forms of alcoholic drinks."

THE following record of a successful endeavor to over-come a morphine habit of several years' growth is abbreviated, by permission of the publishers, from *Lippincott's Magazine* for April, 1868. The absence of the writer in Europe precludes any more definite statement than can be inferred from the narrative itself as to the length of time during which the habit remained uninterrupted. This is a matter of regret, as the *time-element*, in the view of the compiler, enters so largely into the question of the probable recovery of an opium suf-ferer. Morphine appears certainly to have been taken daily in very large quantities for at least five years after the writ-er's habit became established.

Since De Quincey gave to the world his famous " Con-fessions," people have been content to regard opium-eating as a strangely fascinating or as. a strangely horrible vice. Now it is notorious that this practice is on the increase in England, and, as I have recently learned, in this country also. It should be well understood that no man *continues* an opium-eater from choice : he sooner or later becomes the veriest slave ; and it is the object of this paper, originally intended for a friend's hand only, to deter intending neophytes—to warn them from submitting themselves to a yoke which will bow them to the earth. In the hope that it may subserve the good proposed, I venture to give a short account of the experiences of one who still feels in his tissues the yet slowly-smouldering fire of the furnace through which he has passed. I first took opium, in the form of laudanum, nearly ten years ago, for insomnia, or sleeplessness, brought on by overwork at a European university. It seemed as if my tissues lapped up the drug and revelled in the new and strange delight which had opened up to them. All that winter I took doses of from ten to thirty drops every Friday night, there being but few classes on Saturday of any consequence, so that I had the full, uninterrupted effect of the drug. Then I could set to work with unparalleled energy. Thought upon thought flowed to me in never-ending waves. I had a mad striving after intellectual distinction, and felt I would pay any price for it. I generally felt, on the Sunday, my lids slightly heavy, but with a sense pervading me of one who had been taking champagne. I never, however, during this whole winter, took more than one dose a week, varying from thirty to sixty drops. Toward the close of the session I one day deferred the dose till Sunday evening. On the Monday following, in the afternoon, I was in one of the class-rooms listening to the lecturer on Belles-lettres and Rhetoric. One hundred and more young men sat, on that Monday afternoon, listening to his silvery voice as he read extracts from Falconer's " Shipwreck," while the splendid conceptions of the poem, and the opium to boot, taken on the Sunday evening before, were all doing their work on an imaginative young man of nineteen. My blood seemed to make music in my vessels as it seemed to come more highly oxygenized singing to my brain, and tingled fresher and warmer into the capillaries of the entire surface, leaping and bubbling like a mountain-brook after a shower. I knew not at first what it could be, but I felt as if I could have bounded to the desk and taken the place of the professor. For a while, I say, I could not realize the cause. At last, as with a lightning flash, it came. Yes! It was the opium.

And at that moment, then and there was signed the bond which was destined to go far to wither all my fairest hopes; to undermine, while seeming to build up, my highest aspirations; to bring disunion between me and those near and dear to me; to frustrate all my plans, and, while "keeping the word of promise to the ear," ever breaking it to my hope. As I trace these very characters, I am suffering from the remote consequences, in a moral point of view, of having set my hand and seal to that bond.

For two years longer that I remained at college I continued to take laudanum three times a week, and I could, at the end of this period, take two drachms (120 drops) at each dose. All this time my appetite, though not actually destroyed, as it now is, was capricious in the extreme, though I did not lose flesh, at least not markedly so. On the other hand, my capability for mental exertion all through this period was something incredible; and let me say here that one of the most fascinating effects of the drug in the case of an intellectual and educated man is the sense it imparts of what might be termed intellectual daring: add to this the endowments of a strong frame, high animal spirits, and on such an one, opium is the ladder that seems to lead to the gates of heaven. But alas for him when at its topmost rung! After obtaining my degree I gradually eased off the use of the drug for about three months with but little trouble. I was waiting for an appointment in India. At the end of the period named I sailed for my destination, and had almost forgotten the taste of opium; but I found that I was only respited, not redeemed. Two months after I had entered upon my duties, and found myself quietly among my books, the bond was renewed. After two months, in which I passed from laudanum to crude opium, I finally settled on the alka-loid *morphia*, as being the most powerful of all the prepara-tions of opium. I began with half a grain twice a day, and for the six months ending the last day of September of the just expired year, my daily quantum was sixty grains—half taken the instant I awoke, the other half at six o'clock in the evening ; and I could no more have avoided putting into my body this daily supply than I could have walked over a burning ploughshare without scorching my feet.

For the first year, five grains, or even two and a half, would suffice for a couple of days ; that is to say, there was no craving of the system for it during its deprivation for this space. At the end of this period there would be a sense of depression amounting to little beyond uneasiness. But soon four hours' deprivation of the drug gave rise to a phys-ical and mental prostration that no pen can adequately depict, no language convey : a horror unspeakable, a woe unutterable takes possession of the entire being ; a clammy perspiration bedews the surface, the eye is stony and hard, the noise pointed, as in the hippocratic face preceding disso-lution, the hands uncertain, the mind restless, the heart as ashes, the "bones marrowless."

To the opium-consumer, when deprived of this stimulant, there is nothing that life can bestow, not a blessing that man can receive, which would not come to him unheeded, unde-sired, and be a curse to him. There is but one all-absorbing want, one engrossing desire—his whole being has but one tongue—that tongue syllables but one word—*morphia.* And oh ! the vain, vain attempt to break this bondage, the labor worse than useless—a minnow struggling to break the toils that bind a Triton !

I pass over all the horrible physical accompaniments that accumulate after some hours' deprivation of the drug when it has long been indulged in, it being borne in mind that it occurs sooner or later according to the constitution it contends against. Suffice it to say that the tongue feels like a copper bolt, and one seems to carry one's alimentary canal in the brain ; that is to say, one is perpetually reminded that there is such a canal from the constant sense of pain and uneasiness, whereas the perfection of functional performance is obtained when the mind is unconscious of its operation.

The slightest mental or physical exertion is a matter of absolute impossibility. The winding of a watch I have regarded as a task of magnitude when not under the opium influence, and I was no more capable of controlling, under this condition, the cravings of the system for its pabulum, by any exertion of the will, than I, or any one else, could control the dilatation and contraction of the pupils of the eye under the varying conditions of light and darkness. A time arrives when the will is killed absolutely and literally, and at this period you might, with as much reason, tell a man to will not to die under a mortal disease as to resist the call that his whole being makes, in spite of him, for the pabulum on which it has so long been depending for carrying on its work.

When you can with reason ask a man to aerate his lungs with his head submerged in water—when you can expect him to control the movements of his limb while you apply an electric current to its motor nerve—then, but not till then, speak to a confirmed opium-eater of " exerting his will ;" reproach him with want of " determination," and complacently say to him, " Cast it from you and bear the torture for a time." Tell him, too, at the same time, to " do without atmospheric air, to regulate the reflex action of his nervous system and control the pulsations of his heart." Tell the Ethiopian to change his skin, but do not mock the misery and increase the agony of a man who has taken opium for years by talking to him of " will." Let it be understood that after a certain time (varying, of course, according to the capability of physical resistance, mode of life, etc., of the individual) the craving for opium is beyond the domain of the will. So intolerant is the system under a protracted deprivation, that I know of two suicides resulting therefrom. They were cases of Chinese who were under confinement. They were baffled on one occasion in carrying out a previously-successful device for obtaining the drug. The awful mystery of death which they rashly solved had no terrors for them equal to a life without opium, and the morning found them hanging in their cells, glad to get "anywhere, anywhere out of the world."

I have seen another tear his hair, dig his nails into his flesh, and, with a ghastly look of despair and a face from which all hope had fled, and which looked like a bit of shrivelled yellow parchment, implore for it as if for more than life.

But to return to myself. I attained a daily dose of forty grains, and on more than one occasion I have consumed sixty. It became my bane and antidote ; with it I was an *unnatural*—without it, less than man. Food, for months previous to the time of my attaining to such a dose as sixty grains, became literally loathsome ; its sight would sicken me ; my muscles, hitherto firm and well defined, began to diminish in bulk and to lose their contour ; my face looked like a hatchet covered with yellow ochre : and this is the best and truest comparison I can institute. It was sharp, foreshortened and indescribably yellow. I had then been taking *morphia* for nearly two years, but only reached and sustained the maximum doses for the six months already indicated.

Finally, even the sixty grains brought no perceptible increase to the vitality of which the body seemed deprived during its abstinence. It stimulated me to not one-tenth of the degree to which a quarter of a grain had done at the commencement. Still, I had to keep storing it up in me, trying to extract vivacity, energy, life itself, from that which was killing me ; and grudgingly it gave it. I tried hard to free myself, tried again and again ; but I never could at any time sustain the struggle for more than four days at the utmost. At the end of that time I had to yield to my tormentor—yield, broken, baffled, and dismayed—yield to go through the whole struggle over again; forced to poison myself—forced with my own hand to shut the door against hope.

With an almost superhuman effort I roused myself to the determination of doing something, of making one last effort, and, if I failed, to look my fate in the face. What, thought I, was to be the end of all the hopes I once cherished, and which were cherished of and for me by others? of what avail all the learning I had stored up, all the aspirations I nourished?—all being buried in a grave dug by my own hand, and laid aside like funeral trappings, out of sight and memory.

I will not detail my struggles nor speak of the hope which I had to sustain me, and which shone upon me whenever the face of my Maker seemed turned away. Let it suffice that I fought a desperate fight. Again and again I recoiled, baffled and disheartened ; but one aim led me on, and I have come out of the *melée* bruised and broken it may be, but conquering. One month I waged the fight, and I have now been nearly two without looking at the drug. Before, four hours was the longest interval I could endure. Now I am free and the demon is behind me. I must not fail to add that the advantage of a naturally sound and preternaturally vigorous constitution, and (except in the use of opium) one carefully guarded against any of the causes which impart a vicious state of system and so render it incapable of recuperative effort, was my main-stay, and acted the part of a bower-anchor in restoring my general system. This, and a long sea-voyage, aided efforts which would have been otherwise fruitless. On the other hand, let us not too rashly cast a stone at the opium-eater and think of him as a being unworthy of sympathy. If he is not to be envied—as, God knows, he is not—let him not be too much contemned.

I do not now refer to the miserable and grovelling Chinese, who are fed on it almost from the cradle, but to the ordinary cases of educated and intellectual men in this country and in Europe ; and I assert that, could there be a realization of all the aspirations, all the longings after the pure, the good and noble that fill the mind and pervade the heart of a cultivated and refined man who takes to this drug, he would be indeed the paragon of animals. And I go further and say that, given a man of cultivated mind, high moral sentiment, and a keen sense of intellectual enjoyment, blended with strong imaginative powers, and just in proportion as he is so endowed will the difficulty be greater in weaning himself from it. I mean, of course, before the will is killed. When that takes place he is of necessity as powerless as any other victim, and his craving for it is as automatic as in the case of any other opium slave. What he becomes then, I have attempted to describe, and in doing so have suppressed much in consideration of the feelings of those who read.

This it is to be an opium-eater ; and the boldest may well quail at the picture, drawn not by the hand of fancy, but by one who has supped of its horrors to the full, and who has found that the staff on which he leaned has proven a spear which has well-nigh pierced him to the heart. Let no man believe he will escape : the bond matures at last.

L

THE compiler has hesitated as to the propriety of calling attention to the opium-habits of these eminent men, both because little instruction is afforded by the meagre information that is accessible to him respecting their use of opium, and because he apprehends their example may be pleaded in extenuation of the habit. Yet they were confirmed opium-eaters, and remained such to the day of their death; and a reference to their cases may not be without its lesson to that large class of men eminent in public or professional life, who already are, or are in danger of becoming, victims of the opium tyranny, as well as to that larger class who find in undiscriminating denunciations of bad habits, a cheap method of exhibiting a cheap philanthropy.

ROBERT HALL.

With the single exception of Richard Baxter, no clergyman of eminence on record appears to have suffered so acutely or for so long a period from nervous disorders as this eloquent divine. So little, unfortunately, is known of the nature of his disorder, that it would be unjust to express any opinion as to the urgency of the temptation which drove him to the enormous consumption of opium in which he indulged. His biography by Olinthus Gregory sufficiently indicates the severity as well as the early manifestation of his painful disorder. " At about six years of age he was placed at a day-school about

four miles from his father's residence. At first he walked to
school in the morning and home again in the evening. But
the severe pain in his back, from which he suffered so much
through life, had even then begun to distress him; so that he
was often obliged to lie down upon the road; and sometimes
his brother and his other school-fellows carried him in turn.

"Sir James Macintosh described Mr. Hall, when in his
twentieth year, as attracting notice by a most ingenuous and
intelligent countenance, by the liveliness of his manners, and
by such indications of mental activity as could not be misin-
terpreted. His appearance was that of health, yet not of robust health, and he often suffered from paroxysms of pain,
during which he would roll about on the carpet in the utmost
agony; but no sooner had the pain subsided than he would
resume his part in conversation with as much cheerfulness
and vivacity as before he had been thus interrupted.

"At that period, though he was strong and active, he oft-
en suffered extremely from the pain to which I have before
adverted, and which was his sad companion through life.
On entering his room to commence our reading, I could at
once tell whether or not his night had been refreshing; for if
it had, I found him at the table, the books to be studied
ready, and a vacant chair set for me. If his night had been
restless, and the pain still continued, I found him lying on
the sofa, or more frequently upon three chairs, on which he
could obtain an easier position. At such seasons, scarcely
ever did a complaint issue from his lips; but inviting me to
take the sofa, our reading commenced. They, however, who
knew Mr. Hall can conjecture how often, if he became interested, he would raise himself from the chairs, utter a few animated expressions, and then resume the favorite reclining
posture. Sometimes, when he was suffering more than usual, he proposed a walk in the fields, where, with the appropri-

ate book as our companion, we could pursue the subject.
If *he* was the preceptor, as was commonly the case in these
peripatetic lectures, he soon lost the sense of pain, and it
was difficult to say whether the body or the mind were
brought most upon the stretch in keeping up with him.

"During the early months of the year 1803, the pain in
Mr. Hall's back increased both in intenseness and conti-
nuity, depriving him almost always of refreshing sleep, and
depressing his spirits to an unusual degree.

"Often has he been known to sit close at his reading, or
yet more intently engaged in abstract thought, for more than
twelve hours in the day; so that when his friends have called upon him, in the hope of drawing him from his solitude,
they have found him in such a state of nervous excitement
as led them to unite their efforts in persuading him to take
some mild narcotic and retire to rest. The painful result
may be anticipated. This noble mind lost its equilibrium.

"Throughout the whole of Mr. Hall's residence at Leices-
ter, he suffered much from his constitutional complaint; and
neither his habit of smoking nor that of taking laudanum
seemed effectually to alleviate his sufferings. It was truly
surprising that this constant, severe pain, and the means
adopted to mitigate it, did not in any measure diminish his
mental energy.

"In 1812 he took from fifty to one hundred drops every
night. Before 1826 the quantity had increased to one thousand drops.

"Mr. Hall commonly retired to rest a little before eleven
o'clock; but after his first sleep, which lasted about two
hours, he quitted his bed to obtain an easier position on the
floor or upon three chairs, and would then employ himself
in reading the book on which he had been engaged during
the day. Sometimes, indeed often, the laudanum, large as

the doses had become, did not sufficiently neutralize his pain to remove the necessity for again quitting his bed. For more than twenty years he had not been able to pass a whole night in bed. When this is borne in mind it is truly surprising that he wrote and published so much; nay, that he did not sink into dotage before he was fifty years of age.

" Early on the Sunday morning (Mr. Addington says) being requested to see him, I found him in a condition of extreme suffering and distress. The pain in his back had been uncommonly severe during the whole night, and compelled him to multiply at very short intervals the doses of his anodyne, until he had taken no less than 125 grains of solid opium, equal to more than 3000 drops, or nearly four ounces of laudanum ! ! This was the only instance in which I had ever seen him at all overcome by the soporific quality of the medicine ; and it was even then hard to determine whether the effect was owing so much to the quantity administered as to the unusual circumstance of its not having proved, even for a short time, an effectual antagonist to the pain it was expected to relieve.

" The opium having failed to assuage his pain, he was compelled to remain in the horizontal posture ; but while in this situation a violent attack in his chest took place, which in its turn rendered an upright position of the body no less indispensable. The struggle that ensued between these opposing and alike urgent demands became most appalling, and it was difficult to imagine that he could survive it, especially as from the extreme prostration of vital energy, the remedy by which the latter of these affections had often been mitigated—viz., bleeding—could not be resorted to. Powerful stimulants, such as brandy, opium, ether, and ammonia, were the only resources, and in about an hour from my arrival we had the satisfaction of finding him greatly relieved."

The following references to the opium habits of Hall are found in " Gilfillan's Literary Portraits."

" Owing to a pain in his spine, he was obliged to swallow daily great quantities of ether and laudanum, not to speak of his favorite potion, tea. This had the effect of keeping him strung up always to the highest pitch; and, while never intoxicated, he was everlastingly excited. Had he been a feeble man in body and mind the regimen would have totally unnerved him. As it was, it added greatly to the natural brilliance of his conversational powers, although sometimes it appears to have irritated his temper, and to have provoked ebullitions of passion, and hasty, unguarded statements.

" A gentleman in Bradford described to us a day he once spent there with Hall. It was a day of much enjoyment and excitement. At the close of it Hall felt exceedingly exhausted, and on retiring to rest asked the landlady for a wine-glass half full of brandy. ' Now,' he says, ' I am about to take as much laudanum as would kill all this company; for if I don't, I won't sleep one moment.' He filled the glass with strong laudanum, went to bed, and enjoyed a refreshing rest."

JOHN RANDOLPH.

THE eccentricities of no man in America who has been at all conspicuous in public life approach the eccentricities of the late John Randolph of Roanoke. Diseased from his birth, with a temperament of the most excitable kind, he seems during the greater part of his days to have lived only just without the bounds of confirmed insanity. His constitutional infirmities were peculiarly the infirmities that find relief in opium; and it has generally been understood that his addiction to the habit was of many years' continuance and lasted to his death. I have been assured by a Virginia gentleman that when, in one of his last days, he directed his servant to write upon a card for his inspection the word " REMORSE," Randolph was understood to have in mind his excessive use of opium.   His biographer, Mr. Hugh Garland, however, has given apparently as little prominence to his habit in this respect as was consistent with any mention of it whatever.   The letters which follow contain nearly all the information that we can gather from this source.   Under date of February, 1817, Randolph says :

" The worst night that I have had since my indisposition commenced.   It was, I believe, a case of *croup* combined with the affection of the liver and the lungs.   Nor was it unlike tetanus, since the muscles of the neck and back were rigid, and the jaw locked.   I never expected, when the clock struck two, to hear the bell again.   Fortunately, as I found myself going, I dispatched a servant (about one) to the apothecary for an ounce of laudanum.   Some of this, poured down my throat, through my teeth, restored me to something like life. I was quite delirious, but had method in my madness ; for they tell me I ordered Juba to load my gun and to shoot the first ' doctor ' that should enter the room ; adding, ' they are only mustard-seed, and will serve just to sting him.'   Last night I was again very sick ; but the anodyne relieved me. I am now persuaded that I might have saved myself a great deal of suffering by the moderate use of opium."

Under date of March of the same year he writes to a friend : " No mitigation of my worst symptoms took place until the third day of my journey, when I threw physic to the dogs, and instead of opium, etc., I drank, in defiance of my physician's prescription, copiously of cold spring water, and ate plentifully of ice.   Since that change of regimen my strength has increased astonishingly, and I have even gained some flesh, or rather skin."

In a letter to Dr. Brockenbrough, dated May 30, 1828: " I write again to tell you that extremity of suffering has driven me to the use of what I have had a horror all my life—I mean opium—and I have derived more relief from it than I could have anticipated. I took it to mitigate severe pain, and to check the diarrhea. It has done both ; but to my surprise it has had an equally good effect upon my cough, which now does not disturb me in the night, and the diarrhea seldom until toward day-break, and then not over two or three times before breakfast, instead of two or three-and-thirty times.

His biographer, speaking of the state of his health in the autumn of 1831, says, " Mr. Randolph made no secret of his use of opium at this time : ' I live by if not upon opium,' said he to a friend. He had been driven to it as an alleviation of a pain to which few mortals were doomed. He could not now dispense with its use. ' I am fast sinking,' said he, ' into an opium-eating sot, but, please God ! I will shake off the incubus yet before I die ; for whatever difference of opinion may exist on the subject of suicide, there can be none as to *rushing into the presence of our Creator* in a state of drunkenness, whether produced by opium or brandy.' To the deleterious influence of that poisonous drug may be traced many of the aberrations of mind and of conduct so much regretted by his friends during the ensuing winter and spring. But he was by no means under its constant influence."

WILLIAM WILBERFORCE.

So little is known, beyond what appears in the following brief notices, of the opium habits of this distinguished philanthropist, that their citation here would be of little service to opium-eaters, except as they tend to show that the regular use of the drug in small quantities may sometimes be continued for many years without apparent injury to the

health, while the same difficulty in abandoning it is experienced as attends its disuse by those whose moderation has
been less marked.

The son of Wilberforce, in the " Life " of his distinguished
father, says : " His returning health was in a great measure the
effect of a proper use of opium, a remedy to which even Dr.
Pitcairne's judgment could scarcely make him have recourse ;
yet it was to this medicine that he now owed his life, as well
as the comparative vigor of his later years. So sparing was
he always in its use, that as a stimulant he never knew its
power, and as a remedy for his specific weakness he had not
to increase its quantity during the last twenty years he lived.
' If I take,' he would often say, ' but a single glass of wine, I
can feel its effect, but I never know when I have taken my
dose of opium by my feelings.' Its intermission was too
soon perceived by the recurrence of disorder."

In a letter from Dr. Gilman, already quoted in the " Reminiscences of Coleridge," he says, speaking of the difficulty of
leaving off opium, " I had heard of the failure of Mr. Wilberforce's case under an eminent physician of Bath," etc.

A HALF CENTURY'S USE OF OPIUM.

THE case of Wilberforce, however, is thrown into the
shade by that of a gentleman now living in New York, whose
use of opium has been much more protracted than that of
the British philanthropist, and who affirms that opium, instead of weakening his powers of mind or body in any respect, has, on the contrary, been of eminent service to both.
The compiler would have been glad, in the general interests
of humanity, to omit any reference to this case ; but it is a
legitimate part of the story he has undertaken to tell ; and
however this isolated exception to the ordinary results of
the opium habit may be perverted as a snare and delusion

to others, it can not honestly remain untold. In the compiler's interview with this gentleman, now in the one hundred and third year of his age, he was impressed with the
evidences of a physical and mental vigor, and a high moral
tone, which is rarely found in men upon whom rests the
weight of even eighty years. Whatever may be thought of
the convictions of the compiler, as to the enormity of the
injury inflicted upon society from the habitual and increasing
use of opium, he can not reconcile it to his sense of fairness
to omit distinct reference to this most anomalous case.
The gentleman in question was born in England in the
year 1766, and received his first commission in the army in
1786. Serving his country in almost every military station
in the world where the martial drum of England is heard—
in India, at the Cape, in the Canadas, on guard over Napoleon at St. Helena—he illustrates, as almost a solitary exception, the fact that a use of opium for half a century, varying in quantity from forty grains daily to many times this
amount, does not *inevitably* impair bodily health, mental vig-
or, or the higher qualities of the moral nature. The use of
opium was commenced by this gentleman in the year 1816,
as a relief for a severe attack of rheumatism, and has been
continued to the present time, with the exception of a very
brief period when an eminent physician of Berlin, at the suggestion of the late Chevalier Bunsen, the Prussian Embassa-
dor to Great Britain, endeavored to break up the habit. In
this effort he was unsuccessful, and the case remains as a
striking illustration of the weakness of that physiological
reasoning which would deduce certain phenomena as the invariable consequences of a violation of the fundamental
laws of health. Until the chemistry of the living body is
better understood, medical science seems obliged to accept
many anomalies which it can not explain. About all that

can be said of such exceptional cases is this : In the great
conflagrations which at times devastate large cities, some
huge mass of solid masonry is occasionally seen in the midst
of the wide-spread ruin, looking down upon prostrate col-
umns, broken capitals, shattered walls, and the cinders and
ashes of a general desolation. The solitary tower unquestionably stands ; but its chief utility lies in-this, that it serves
as a striking monument of the appalling and wide-spread destruction to which it is the sole and conspicuous exception.

MOST of the preceding pages were already prepared for the press, when the attention of the compiler was attracted by a very remarkable article in *Harper's Magazine* for August, 1867, entitled, "What Shall They Do to be Saved?" The graphic vividness of the story, as well as the profound insight and wide experience with which it was written, led me to solicit from the unknown author the addition of it to the pages of my own book. It proved to be from the pen of Fitz Hugh Ludlow, already recognized by the public as a writer of eminence, both in science and letters. The permission being freely accorded, I was still further moved to ask that he would give me a statement of the method pursued by him in dealing with the class to which it refers. The letter following his article was his response to my request. It will be seen to contain an outline of his views upon the subject to which he has devoted some years of study and practice, and is especially valuable as embodying the germ of a plan by which, according to his growing conviction, the opium-eater can alone be saved. As the conclusions of a writer who seems to the compiler to be singularly intelligent and definite in his knowledge of this most interesting and difficult field of disease and treatment, it needs no further recommendation to the attention of the reader. Since the publication of his August article, a multitude of letters received from all portions of the country, asking his advice and assistance in such cases as this book describes, has left a profound conviction upon his mind of the most crying need of the establishment of an institution where opium-eaters can be treated specially. In this view of the urgent necessities of the case, the compiler most heartily and earnestly concurs.

———

I have just returned from forty-eight hours' friendly and professional attendance at a bedside where I would fain place every young person in this country for a single hour before the Responsibilities of Life have become the sentinels and Habit the jailer of his Will.

My patient was a gentleman of forty, who for several years of his youth occasionally used opium, and for the last eight has habitually taken it. During these eight years he has made at least three efforts to leave it off, in each instance diminishing his dose gradually for a month before its entire abandonment, and in the most successful one holding the enemy at bay for but a single summer. In two cases he had no respite of agony from the moment he dropped till he resumed it. In the third case, a short period of comparative repose succeeded the first fiery battle, but in the midst of felicitations on his victory he was attacked by the most agonizing hemicranial headaches (resulting from what I now fear to have been already permanent disorganization of the stomach), and went back to his nepenthe in a state of almost suicidal despair, only after the torture had continued for weeks without a moment's mitigation.

He had first learned its seductions, as happens with the vast majority of Anglo-Saxon opium-eaters, through a medical prescription. An attack of inflamed cornea was treated with caustic applications, and the pain assuaged by internal doses of M'Munn's Elixir. When my friend came out of his dark room and bandages at the end of a month he had consumed twenty ounces of this preparation, whose probable distinction from the tincture known as laudanum I point out

below in the note.* Here it may not be superfluous to say
that the former preparation has all the essential properties
of the latter, save certain of the constipatory and stupefying
tendencies which, by a private process known to the assigns
of the inventor, have been so masked or removed that it possesses in many cases an availableness which the practitioner
can not despise, though compelled by the secrecy of its formula to rank it among quack medicines. The amount of it
which my friend had taken during his month's eclipse represents an ounce of dry gum opium—in rough measurement a
piece as large as a French billiard ball. I thus particular-
ize because he had never previously been addicted to the
drug ; had inherited a sound constitution, and differed from
any other fresh subject only in the intensity of his nervous
temperament. I wish to emphasize the fact that the system
of a mere neophyte, with nothing to neutralize the effects of
the drug save the absorbency, so to speak, of the pain for
which it was given, could so rapidly adapt itself to them as
to demand an increase of the dose in such an alarming ratio.
There are certain men to whom opium is as fire to tow, and
my friend was one of these. On the first of October he sensibly perceived the trifling dose of fifty drops ; on the first
of November he was taking, without increased sensation, an
ounce vial of " M'Munn " daily.

* Mr. Frank A. Schlitz has kindly made for me a special analysis of
M'Munn's Elixir, which seems to prove that the process of its preparation
amounts to more than the *denarcotization* of opium, which is spoken of on
the wrapper of each vial. As nearly as can be ascertained, M'Munn's
Elixir is simply an aqueous infusion of opium—procured by the ordinary
maceration—and preserved from decomposing by the subsequent addition
of a small portion of alcohol. *Narcotin* being absolutely insoluble in water
is eliminated as the circular says. This fact alone would not account for the
difference between its action and that of laudanum. This is explained by
the fact that all the other alkaloids possess diverse rates of solubility in
water, and exist in M'Munn's Elixir in very different relative proportions

From that time — totally ignorant of the terrible trap which lay grinning under the bait he dabbled with—he continued to take opium at short intervals for several years. When by the physician's orders he abandoned " M'Munn," on the subsidence of the eye-difficulty, his symptoms were uneasy rather than distressing, and disappeared after a few days' oppression at the pit of the stomach and a few nights' troubled dreaming. But he had not forgotten the sweet dissolving views at midnight, the great executive achievements at noonday, the heavenly sense of a self-reliance which dare go anywhere, say any thing, attempt any thing in the world. He had not forgotten the nonchalance under slight, the serenity in pain, the apathy to sorrow, which for one month set him calm as Boodh in the temple-splendors of his darkened room. He had not forgotten that the only perfect *peace* he had ever experienced was there, and he remembered that peace as something which seemed to blend all the assuaged passion and confirmed dignity of old age with that energy of high emprise which thrills the nerves of manhood. He had tasted as many sources of earthly pleasure as any man I ever knew ; but the ecstasies of form and color, wine, Eros, music, perfume, all the luxuries of surrounding which wealth could purchase or high-breeding appreciate, were as nothing to him in comparison with the memory of that time on which his family threw away their sympathy when they called it his " month of *suffering.*"

Accordingly, without much more instinct of concealment than if it were an occasional tendency to some slight convivial excess, he had resort to M'Munn, in ounce doses, whenever the world went wrong with him. If he had a headache or a toothache ; if the weather depressed him ; if he had a cer-

from those which they bear to each other in the alcoholic tincture called laudanum.

tain "stint" of work to do without the sense of native vigor to accomplish it ; if he was perplexed and wished to clear his head of passion ; if anxieties kept him awake ; if irregularities disturbed his digestion—he had always one refuge certain. Nŏ fateful contingency could pursue him inside M'Munn's enchanted circle. He was a young and wealthy bachelor, living the life of a refined *bon vivant;* an insatiable traveller, surrounded by flatterers, and without a single friend who loved him enough to warn him of his danger excepting those who, like himself, were too ignorant to know it. After three years of dalliance he became an habitual user of opium, and had been one for eight years when I was first called to him.

By the time that the daily habit fastened itself he had learned of other opiate preparations than M'Munn's, and finding a certain insufficiency characterize that tincture as he increased the size of the dose, had recourse to laudanum, which contains the full native vigor of the drug unmodified. This nauseated him. He had the same experience with gum opium, opium pills, and opium powder ; so that he was driven to that form of exhibition which sooner or later naturally strikes almost every opium-eater as the most portable, energetic, and instantaneous—morphia or óne of its salts. My friend usually kept the simple alkaloid in a paper, and dissolved it as he needed it in clear water, sometimes substituting an equivalent of "*Magendie's Solution*," which contains sixteen grains of the salt diffused through an ounce of water by the addition of a few drops of sulphuric acid. When I first saw him he had reached a daily dose of twelve grains of sulphate of morphia, and on occasions of high excitement had increased his dose without exaggerating the sensible effect to nearly twenty. The twelve which formed his habitual *per diem* were divided into two equal doses, one taken immediately after rising, the other just about sun-down.

As yet he had not begun to feel the worst physical effects which sooner or later visit the opium-eater. His digestion seemed unimpaired so long as he took his morphia regular-ly; he was sallow and somewhat haggard, but thus far no distressing biliary symptoms had manifested themselves; his sleep was always dreamy, and he woke at short intervals during the night, but invariably slept again at once, and had so adjusted himself to the habit as to show no signs of suf-fering from wakefulness; his hand was steady; his muscular system easily exhausted, but by no means what one would call feeble. As he himself told me, he had come to the con-clusion to emancipate himself because opium-eating was a horrible mental bondage. The physical power of the drug over him he only realized when attempting its abandonment. Its spiritual thraldom was his hourly misery. He was con-nected by blood and marriage with several of the best fam-ilies in the land. Money had not been stinted in his educa-tion, and his capabilities were as great as his advantages. He was one of the bravest, fairest, most generous natures I ever came in contact with; was versatile as a Yankee Crich-ton; had ridden his own horse in a trotting match and beat-en Bill Woodruff; had carried his own little 30-ton schooner from the Chesapeake to the Golden Gate through the Straits of Magellan; had swum with the Navigators' Islanders, shot buffalo, hunted chamois, and lunched on mangosteens at Pe-nang. Through all his wanderings the loftiest sense of what was heroic in human nature and divine in its purified form, the monitions of a most tender conscience, and the echoes of that Puritan education which above all other schemes of training makes human responsibility terrible, had gone with him like his tissue. He saw the good and great things within reach of a fulfilled manhood, and of a sudden waked up to feel that they could on earth never be his. He was naturally very truthful, and, although the invariable tendency of opium-eaters is to extirpate this quality, could not flatter himself. Other minds around him responded to a sudden call as his own did not. Every day the need of energy took him more by surprise.

The image-graving and project-building characteristic of opium, which comes on with a sense of genial radiation from the epigastrium about a quarter of an hour after the dose, had not yet so entirely disappeared from its effect on him, as it always does at a later stage of the indulgence. But instead of being an instigation to the delightful reveries which ensued on his earlier doses, this peculiarity was now an executioner's knout in the hands of Remorse. He was daily and nightly haunted by plans and pictures whose feverish unreal beauty he remembered having seen through a hundred times. Those Fata Morgana plans, should he again waste on them the effort of construction? The result had been a chaos of aimless, ineffectual days. Those pictures, why were they brought again to mock him? Were they not horrible impossibilities? Were they not, through the paralysis of his executive faculties, mere startling likenesses of Disappointment? In his opium dreams he had seen his own ships on the sea; commerce bustling in his warehouse; money overflowing in his bank; babies crowing on his knee; a wife nestling at his breast; a basso voice of tremendous natural power and depth scientifically cultivated to its utmost power of pleasing artists or friends; a country estate on the Hudson, or at Newport, with emerald lawns sloping down to the amber river or the leek-green sea; the political and social influence of a great landholder. How pleasurably he had once perceived all these possible joys and powers! How undeludedly he now saw their impossible execution!

So, coming to me, he told me that his object in trying to leave off opium was to escape from these horrible ghosts of a life's unfulfilled promise. Only when he tried to abandon opium did he realize the physical hold the drug had on him. Its spiritual thraldom was his hourly misery.

For three months I tried to treat him in his own house, here in the city. A practitioner of any experience need not be told with what success. I could reduce him to a dose of half a grain of sulphate of morphia a day, keep him there one week, and making a morning call at the expiration of that time discover that some nocturnal nervous paroxysm had necessitated either a return to five grains or a use of brandy (which, though no drinker, he tried to substitute) sufficient to demand a much larger dose of opium in its reaction. He had lost most of his near connections, and not for one hour could any hired attendant have withstood his appeal, or that marvellous ingenuity by which, without appeal, the opium-eater obtains the drug which, to him, is like oxygen to the normal man.

This ingenuity manifests itself in subterfuges of a complicated construction and artistic plausibility which might have puzzled Richelieu; but it is really nothing to wonder at when we recollect the law of nature by which any extreme agony, so long as it continues remediable, sharpens and concentrates all a man's faculties upon the one single object of procuring the remedy. If my house is on fire, I run to the hydrant by a mere automatic operation of my nerves. If my leg is caught in the bight of a paying-out hawser, my whole brain focuses at once on that single thought, "*an axe.*" If I am enduring the agony which opium alone can cause and cure, every faculty of my mind is called to the aid of the tortured body which wants it. When a man has used opium for a long time the condition of brain supervening on his deprivation of the drug for a period of twenty-four hours is such as very frequently to render him suicidal. Cottle tells us how Coleridge one day took a walk along Bristol wharves, and sent his attendant down the pier to inquire the name of a vessel, while he slipped into a druggist's on the quay and bought a quart of laudanum ; but in no fibre of his nature could Cottle conceive the awful sense of a force despotizing it over his will, a degradation descending on his manhood, which Coleridge felt as he concentrated on that one single cry of his animal nature and the laudanum which it spoke for, all the faculties of construction and insight which had created the " Ancient Mariner " and the " Aids to Reflection."

Likewise I suppose there are very few people who could patiently regard the fact that one of the very purest and bravest souls I ever knew had become so demoralized by the perseverance of disease and suffering as to deal like a lawyer with his best friends, and shuffle to the very edge of falsehood, when his nature clamored for opium. I was particular to tell him whenever I detected any evasion (an occasion on which his shame and remorse were terrible to witness) that *I*, personally, had none the less respect for him. I knew he was dominated, and in no sense more responsible for breaking his resolution than he would have been had he vowed to hold his finger in the gas-blaze until it burned off. In this latter case the mere translation of chemical decomposition into pain, and round the automatic nerve-arc into involuntary motion, would have drawn his finger out of the blaze, as it did in the cases of Mutius Scævola and Cranmer, if they ever attempted the feat credited them by tradition. In his case the abandonment of opium brought on an agony which took his actions entirely out of voluntary control, eclipsing the higher ideals and heroisms of his imagination at once, and reducing him to that automatic condition in which the nervous system issues and enforces only those edicts which are counselled by pure animal self-preservation. Whatever may have been the patient's responsibility in *beginning* the use of narcotics or stimulants (and I usually find, in the case of opium-eaters, that its degree has been very small indeed, therapeutic use often fixing the habit forever before a patient has convalesced far enough even to know what he is taking) habituation invariably tends to reduce the man to the *automatic* plane, in which the will returns wholly to the tutelage of sensation and emotion, as it was in infancy; while all the Intellectual, save *Memory*, and the most noble and imperishable among the Moral faculties may survive this disorganization for years, standing erect above the remainder of a personality defrauded of its completion to show what a great and beautiful house might have been built on such strong and shapely pillars. Inebriates have been repeatedly known to risk imminent death if they could not reach their liquor in any other way. The grasp with which liquor holds a man when it turns on him, even after he has abused it for a lifetime, compared with the ascendency possessed by opium over the unfortunate habituated to it for but a single year, is as the clutch of an angry woman to the embrace of Victor Hugo's *Pieuvre*. A patient whom, after habitual use of opium for ten years, I met when he had spent eight years more in reducing his daily dose to half a grain of morphia, with a view to its eventual complete abandonment, once spoke to me in these words :

"God seems to help a man in getting out of every difficulty but opium. There you have to *claw* your way out over red-hot coals on your hands and knees, and drag yourself by main strength through the burning dungeon-bars."

This statement does not exaggerate the feeling of many another opium-eater whom I have known.

Now, *such* a man is a proper subject, not for *reproof*, but for *medical treatment.* The problem of his case need embarrass nobody. It is as purely physical as one of small-pox. When this truth is as widely understood among the laity as it is known by physicians, some progress may be made in staying the frightful ravages of opium among the present generation. Now, indeed, it is a difficult thing to prevent relatives from exacerbating the disorder and the pain of a patient, who, from their uninformed stand-point, seems as sane and responsible as themselves, by reproaches at which they would shudder, as at any other cruelty, could they be brought to realize that their friend is suffering under a disease of the very machinery of volition ; and no more to be judged harshly for his acts than a wound for suppurating or the bowels for continuing the peristaltic motion.

Finding—as in common with all physicians I have found so many times before—that no control of the case could be obtained while the patient stayed at home, and deeply renewing my often-experienced regret that the science and Christian charity of this country have perfected no scheme by which either inebriates or opium-eaters may be properly treated in a special institution of their own, I was at length reluctantly compelled to send my friend to an ordinary water-cure at some distance from town.

The cause of my reluctance was not the prospect of a too liberal use of water, for by arrangement with the heads of the establishment I was able to control that as I chose ; moreover, an employment of the hot-bath in what would ordinarily be excess is absolutely necessary as a sedative throughout the first week of the struggle. I have had several patients whom during this period I plunged into water

at 110° Fahrenheit as often as fifteen times in a single day—
each bath lasting as long as the patient experienced relief.
In some cases this Elysium coming after the rack has been
the only period for a month in which the sufferer had any
thing resembling a doze. My reluctance arose from the necessity of sending a patient in such an advanced stage of the
opium disease so far away from me that I must rely on re-
ports written by people without my eyes, for keeping personally *au courant* with the case ; that I must consult and pre-
scribe by letter, subject to the execution of my plans by men,
who, though excellent and careful, were ignorant of my theories of treatment, and had never made this particular disease a specialty. I accordingly sent Mr. A. away to the
water-cure, all friendless and alone to fight the final battle
of his life against tougher odds than he had ever before encountered. At no time in my life have I realized with greater bitterness the helplessness of a practitioner who has no
institution of his own to take such cases to than when I
shook his poor, dry, sallow hand and bade him good-bye at
the station.

As I said in the beginning, I am just home from seeing
the result. Mr. A. has fared as special cases always do in
places where there is no special provision for them. To
speak plainly, he had been badly neglected ; and that, undoubtedly, without the slightest intention on the part of the
heads of the house to do other than their duty. Six weeks
ago I heard from the first physician that my friend was entirely free from opium, and, though still suffering, was steadily on the mend. I had no further news from him till I was
called to his bedside by a note which said he feared he was
dying, pencilled in a hand as tremulously illegible as the confession of Guy Fawkes. I was with him by the earliest
train I could take, after arranging with a neighbor for my

practice, and found him in a condition which led him to say,
as I myself said at the commencement of this article:
"Would to God that every young person could stand for a
single hour by this bedside before Life's Responsibilities
have become the sentinels and Habit the jailer of the Will!"

I had not been intelligently informed respecting the progress of his case. He had been better at no time when I was
told he was so, though his freedom from opium had been of
even longer duration than I was advised. *For ninety days
he had been without opium in any form.* The scope of so untechnical an article leaves no room to detail what had been
done for him as alleviation. His prostration had been so
great that he could not correspond with me himself until the
moment of his absolute extremity; and only after repeated
entreaties to telegraph to myself and his family had been refused on the ground that his condition was not critical, he
managed to get off the poor scrawl which brought me to his
side.

For the ninety days he had been going without opium he
had known nothing like proper sleep. I desire to be understood with mathematical literalness. There had been periods when he had been *semi-conscious;* when the outline of
things in his room grew vaguer and for five minutes he had
a dull sensation of not knowing where he was. This temporary numbness was the only state which in all that time simulated sleep. From the hour he first refused his craving, and
went to the battle-field of bed, he had endured such agony as
I believe no man but the opium-eater has ever known. I am
led to believe that the records of fatal lesion, mechanical
childbirth, cancerous affection, the stake itself, contain no
greater torture than a confirmed opium-eater experiences in
getting free. Popularly this suffering is supposed to be
purely intellectual—but nothing can be wider of the truth.

Its intellectual part is bad enough, but the physical symptoms are appalling beyond representation. The look on the face of the opium sufferer is indeed one of such keen mental anguish that outsiders may well be excused for supposing that is all. I shall never forget till my dying-day that awful Chinese face which actually made me rein my horse at the door of the opium *hong* where it appeared, after a night's debauch, at six o'clock one morning when I was riding in the outskirts of a Pacific city. It spoke of such a nameless horror in its owner's soul that I made the sign for a pipe and proposed, in "*pigeon English,*" to furnish the necessary coin. The Chinaman sank down on the steps of the *hong,* like a man hearing medicine proposed to him when he was gangrened from head to foot, and made a gesture, palms downward, toward the ground, as one who said, " It has done its last for me—I am paying the matured bills of penalty." The man had exhausted all that opium could give him ; and now, flattery past, the strong one kept his goods in peace. When the most powerful alleviative known to medical science has bestowed the last Judas kiss which is necessary to emasculate its victim, and, sure of the prey, substitutes stabbing for blandishment, what alleviative, stronger than the strongest, shall soothe such doom? I may give chloroform. I always do in the *dénouement* of bad cases—ether—nitrous oxyd. In employing the first two agents I secure rest, but I induce death nine cases out of ten. Nothing is better known to medical men than the intolerance of the system to chloroform or ether after opium. Nitrous oxyd I am still experimenting with, but its simple undiffused form is too powerful an agent to use with a patient who for many days must be hourly treated for persevering pain. So the opium-eater is left as entirely without anæsthetic as the usual practice leaves him without therapeutic means. Both here and abroad opium-

M

eaters have discovered the fact that, in an inveterate case, where opium fails to act on the brain through the exhausted tissues of the stomach, bichlorid of mercury in combination with the dose behaves like a *mordant* in the presence of a dye, and, so to speak, *precipitates* opium upon the calloused surfaces of the mucous and nervous layers. This expedient soon exhausts itself in a death from colliquative diarrhea, produced partly by the final decompositions of tissue which the poisonously antiseptic property of opium has all along improperly stored away ; partly by the definite corrosions of the new addition to the dose. But in no case is there any relief to a desperate case of opium-eating save death.

Remembering that Chinaman's face, I can not wonder at the popular notion regarding the abandonment of opium. Men say it is a mental pain ; because spiritual woe is the expression of the sufferer's countenance. And so it is, but this woe is underlain by the keenest brute suffering. Let me sketch the opium-eater's experience on the rugged road upward.

Let us suppose him a resolute man, who means to be free, and with that intent has reduced to a hundred drops the daily dose which for several years had amounted to an ounce of laudanum. I am not supposing an extreme case. An ounce of laudanum is a small *per diem* for any man who has taken his regular rations of the drug for a twelvemonth. In the majority of cases I have found an old *habitué's* daily portion to exceed three, or the equivalent of that dose in crude opium or morphia ; making seventy-two grains of the gum or twelve of its most essential alkaloid. In one most interesting case I found a man who having begun on the first of January with one half a grain of sulphate of morphia for disease, at the end of March was, to all appearance, as hopeless an opium-eater as ever lived, taking thirty-two grains of the

salt per day in the form of *Magendie's Solution.* This, however, was an unusual case. According to my experience
the average opium-eater reaches twelve grains of morphia in
ten years, and may live after that to treble the amount: the
worst case I ever knew attaining a dose of ninety grains, or
one and a half of the drachm vials ordinarily sold. I am happy, in passing, to add that for more than two years both the
extreme cases just mentioned have been entirely cured.

If the opium-eater has been in the habit of dividing his
daily dose he begins to feel some uneasiness within an hour
after his first deprivation, but it amounts to nothing more
than an indefinite restlessness. In any case his first well-
marked opium torments occur early after he has been without the drug for twenty-four hours.

At the expiration of that time he begins to feel a peculiar
corded and *tympanic* tightness about the epigastrium. A
feverish condition of the brain, which sometimes amounts to
absolute *phantasia,* now ensues, marked off into periods of increasing excitement by a heavy sleep, which, after each interval, grows fuller of tremendous dreams, and breaks up
with a more intensely irritable waking. I have held a man's
hand while he lay dreaming about the thirty-sixth hour of his
struggle. His eyes were closed for less than a minute by
the watch, but he awoke in a horrible agony of fear from
what seemed to have been a year-long siege of some colossal
and demoniac Vicksburg.

After the opium-eater has been for forty-eight hours without his solace this heavy sleep entirely disappears. While it
stays it never lasts over half an hour at a time, and is so
broken by the crash of stupendous visions as not to amount
to proper slumber. During its period of continuance the
opium-eater woos its approaches with an agony which shows
his instinct of the coming weeks of sleeplessness. It never

rests him in any valid sense. It is a congestive decomposition rather than any normal reconstruction of the brain. He wakes out of it each time with a heart more palpitating ; in a perspiration more profuse ; with a greater uncertainty of sense and will ; with a more confused memory ; in an intenser agony of body and horror of hopelessness.

Every nerve in the entire frame now suddenly awakes with such a spasm of revivification that no parallel agony to that of the opium-eater at this stage can be adduced, unless it be that of the drowned person resuscitated by artificial means. Nor does this parallel fully represent the suffering, for the man resuscitated from drowning re-oxydizes all *his* surplus carbon in a few minutes of intense torture, while the anguish which burns away that carbon and other matter, properly effete, stored away in the tissues by opium, must last for hours, days, and weeks. Who is sufficient for this long, *long* pull ?

From the hour this pain begins to manifest itself it continues (in any average case of a year's previous habituation to the drug) for at least a week without one second's lull or exhaustion. A man may catch himself dozing between spasms of tic-douloureux or toothache ; he never doubts whether he is awake one instant in the first week after dropping his opium. One patient whom I found years ago at a water-cure followed the watchman all night on crutches through his tour of inspection around the establishment. Other people, after walking a long time, shift from chair to chair in their rooms, talking to any body who may happen to be present in a low-voiced suicidal manner, which inexperience finds absolutely blood-freezing. Later such rock to and fro, moaning with agony, for hours at a time, but saying nothing. Still others go to their beds at once, and lie writhing there until the struggle is entirely decided. I have learned that this last class is generally the most hopeful.

The period during which this pain is to continue depends upon two elements.

1st. How long has the patient habitually taken opium?

2d. How much constitutional strength remains to throw it off?

" How much has he taken in the aggregate ?" is practically not an equivalent of the first question. I have found an absolutely incurable opium-eater who had never used more than ten grains of morphia *per diem;* but he had been taking it habitually for a dozen years. In another case the patient had for six months repeated before each meal the ten-grain dose which served the other all day ; but he was a man whose pluck under pain equalled that of a woman's, and after a fortnight's anguish of such horror that one could scarcely witness it without being moved to tears, came out into perfect freedom. The former patient, although he had never in any one day experienced such powerful effects from opium as the latter, had used the drug so long that every part of his system had reconstructed itself to meet the abnormal conditions, and must go through a second process of reconstruction, without any anodyne to mask the pain resulting from its decomposition, before it could again tolerate existence of the normal kind. If opium were not an anodyne the terrible structural changes which its works would cause no surprise ; it would be *felt* eating out its victim's life like so much nitric acid. During the early part of the opium-eater's career these structural changes go on with a rapidity which partly accounts for the vast disengagements of nervous force, the exhilaration, the endurance of effort, which characterize this stage, later to be substituted by utter nervous apathy. By the time the substitution occurs something has taken place

throughout the physical structure which may be rudely liken-
ed to the final equilibrium of a neutral salt after the efferves-
cence between an acid and an alkali.   So to speak, the tissues have now combined with their full equivalent of all the
poisonous alkaloids in opium.   Further use of it produces no
new disengagements of nervous force ; the victim may double,
quadruple his dose, but he might as well expect further ebul-
lition by adding more aqua-fortis to a satisfied nitrate as to
develop with opium exhilarating currents in a tissue whose
combination with that drug have already reached their chemical limit.*

The opium-eater now only continues his habit to preserve
the terrible static condition to which it has reduced him, and
to prevent that yet more terrible dynamic condition into
which he comes with every disturbance of equilibrium ; a
condition of energetic and agonizing dissolutions which must
last until every fibre of wrongly-changed tissue is burned
up and healthily replaced.   Though I have called the early
reactions of opium rapid, they are necessarily much less so
than those produced by a simple chemical agent.   No drug
approaches it in the possession of *cumulative* characteristics ;
its dependence on the time element must therefore be always
carefully considered in treating a case.   This fact leads us
to understand the other .element in the question, how long .
the torments of the opium-fighter must continue.   Having
ascertained the chronology of his case, we must say, " Given
this period of subjection, has the patient enough constitution-

---

* I say " chemical " because so much it is possible to know experimentally ; and the very interesting examination of such higher forces as constantly seem to intrude in any nervous disturbance would here involve
the discussion of a theoretical " vital principle "— something apart from
and between the soul and physical activities—which scientific men are
universally abandoning.

al vigor left to endure the period of reconstruction which must correspond to it?" *

I am naturally sanguine, and began my study of opium-eaters with the belief that none of them were hopeless. Experience has taught me that there is a point beyond which any constitution—especially one so abnormally sensitive as the opium-eater's—can not endure keen physical suffering without death from spinal exhaustion. I once heard the eminent Dr. Stevens say that he made it a rule never to attempt a surgical operation if it must consume more than an hour. Similarly, I have come to the conclusion never to amputate a man from his opium-self if the agony must last longer than three months. Uneasiness, corresponding to the irritations of dressing a stump—may continue a year longer; a few victims of the habit outlive a certain opium-prurience, which has also its analogue in the occasional titillation of a healed wound—these are comparatively tolerable; but, if we expect to save a patient's life, we must not protract an agony which so absolutely interferes with normal sleep as that of the opium-eater's for longer than three months in the case of any constitution I have thus far encountered.

Usually as early as the third day after its abandonment (unless the constitution has become so impaired by long habituation that there will probably be no vital reaction) opium begins to show its dissolutions from the tissue by a profuse and increasingly acrid bilious diarrhea, which must not be checked if diagnosis has revealed sufficient constitutional vigor to justify any attempt at abandonment of the

* Not correspond day by day. At that rate a reforming opium-eater (I use the participle in the *physical* sense, for very few opium-eaters are more to blame than any other sick persons) must pay a "shent per shent" which no constitution could survive. The correspondence is simply proportional.

drug. Hemorrhoids may result; they must be topically treated; mild astringents may be used when the tendency seems getting out of eventual control; bland foods must be given as often as the usually fastidious appetite will tolerate them; the only tonic must be beef-tea—diffusible stimulus invariably increasing the agony, whether in the form of ale, wine, or spirits. Short of threatened collapse, the bowels must not be retarded. There is nothing in the faintest degree resembling a substitute for opium, but from time to time various alleviatives, which can not be discussed in an untechnical article, may be administered with benefit. The spontaneous termination of the diarrhea will indicate that the effete matters we must remove have been mainly eliminated, and that we may shortly look for a marked mitigation of the pain, followed by conditions of great debility but increasingly favorable to the process of reconstruction. That process, yet more than the alleviate, demands a book rather than an article.

I have intentionally deferred any description of the agony of the opium struggle, as a *sensation*, until I returned from depicting general symptoms, to relate the particular case which is my text. The sufferings of the patient, from whom I have just returned, are so comprehensive as almost to be exhaustively typical.

When simple nervous excitement had for two days alternated with the already mentioned intervals of delirious slumber, a dull, aching sensation began manifesting itself between his shoulders and in the region of the loins. Appetite for food had been failing since the first denial of that for opium. The most intense gastric irritability now appeared in the form of an aggravation of the tympanic tightness, corrosive acid ructations, heart-burn, water-brash, and a peculiar sensation, as painful as it is indescribable, of *self-consciousness* in the whole upper part of the digestive canal. The best idea of this last symptom may be found by supposing all the nerves of involuntary motion which supply that tract with vitality, suddenly to be gifted with the exquisite sensitiveness to their own processes which is produced by its correlative object in some organ of special sense—the whole organism assimilating itself to a retina or a finger-tip. Sleep now disappeared. This initiated an entire month during which the patient had not one moment of even partial unconsciousness.

In less than a week from the beginning the symptoms indicated a most obstinate chronic gastritis. There was a perpetual sense of corrosion at the pit of the stomach very like that which characterizes the fatal operation of arsenic. There was less action of the liver than usually indicates a salvable case, and no irritation of the lowest intestines. *Pari passu* with the gastritic suffering, the neuralgic pain spread down the extremities from an apparent centre between the kidneys, through the trunk, from another line near the left margin of the liver, and through the whole medullary substance of the brain itself. Although I was so unfortunate as not to be beside him during this stage, I can still infallibly draw on my whole experience for information regarding the intensity of this pain. *Tic-douloureux* most nearly resembles it in character. Like that agonizing affection, it has periods of exacerbation ; unlike it, it has no intervals of continuous repose. Like *tic-douloureux*, its sensation is a curiously fluctuating one, as if pain had been *fluidized* and poured in trickling streams through the tubules of nerve tissue which are affected by it ; but, unlike that, it affects every tubule in the human body—not a single diseased locality. Charles Reade chaffs the doctors very wittily in "Hard Cash" on their *penchant* for the word " *hyperæsthesia*," but nothing else exactly defines that exaggeration of nervous sensibility which

M 2

I have invariably seen in opium-eaters. Some of them were hurt by an abrupt slight touch, and cried out at the jar of a heavy footstep like a patient with acute rheumatism. Some developed sensitiveness with the progress of expurgating the poison, until their very hair and nails felt sore, and the whole surface of the skin suffered from cold air or water like the lips of a wound. After all, utterly unable to convey an idea of the *kind* of suffering, I must content myself by repeating, of its extent, that no prolonged pain of any kind known to science can equal it. The totality of the experience is only conceivable by adding this physical torture to a mental anguish which even the Oriental pencil of De Quincey has but feebly painted ; an anguish which slays the will, yet leaves the soul conscious of its murder ; which utterly blots out hope, and either paralyzes the reasoning faculties which might suggest encouragements, or deadens the emotional nature to them as thoroughly as if they were not perceived ; an anguish which sometimes includes just, but always a vast amount of *unjust* self-reproach, which brings every failure and inconsistency, every misfortune or sin of a man's life as clearly before his face as on the day he was first mortified or degraded by it—before his face, not in one terrible dream, which is once for all over with sunrise, but as haunting ghosts, made out by the feverish eyes of the soul down to the minutest detail of ghastliness, and never leaving the side of the rack on which he lies for a moment of dark or day-light, till sleep, at the end of a month, first drops out of heaven on his agony.

A third element in the suffering must briefly be mentioned. It results directly from the others. It is that exhaustion of nervous power which invariably ensues on protracted pain of mind or body. It proceeds beyond reaction to collapse in a hopeless case ; it stops this side of that in a salvable one.

On reaching his room I found my friend bolstered upright in bed, with a small two-legged crutch at hand to prop his head on when he became weary of the perpendicular position. This had been his attitude for fifty days. Whether from its impeding his circulation, the distribution of his nervous currents, or both, the prostrate posture invariably brought on cessation of the heart and the sense of intolerable strangling. His note told me he was dying of heart disease, but, as I expected, I found that malady merely simulated by nervous symptoms, and the trouble purely functional. His food was arrow-root or sago, and beef-tea. Of the vegetable preparation he took perhaps half a dozen table-spoonfuls daily; of the animal variable quantities, averaging half a pint per diem. This, though small, was far from the minimum of nutriment upon which life has been supported through the most critical periods. Indeed, I have known three patients tided over stages of disease otherwise desperately typhoid by beef-tea baths, in which the proportion of ozmazone was just perceptible, and the sole absorbing agency was a faint activity left in the pores of the skin. But these patients had suffered no absolute disorganization. The practitioner had to encounter a swift specific poison, not to make over tissues abnormally misconstructed by its long insidious action. On examination I discovered facts which I had often feared, but never before absolutely recognized, in my friend's case. The stomach itself, in its most irreproducible tissue, had undergone a partial but permanent disorganization. The substance of the organ itself had been altered in a way for which science knows no remedy.

Hereafter, then, it can only be rechanged by that ultimate decomposition which men call death. Over the opium-eater's coffin at least, thank God! a wife and a sister can stop weeping and say, " He's free."

I called to my friend's bedside a consultation of three physicians and the most nearly related survivor of his family. I laid the case before them ; assisted them to a full *prognosis ;* and invited their views. I spent two nights with my friend. I have said that during the first month of trial he had not a moment of even partial unconsciousness. Since that time there had been perhaps ten occasions a day, when for a period from one minute in length to five, his poor, pain-wrinkled forehead sank on his crutch, his eyes fell shut, and to outsiders he seemed asleep. But that which appeared sleep was internally to him only one stupendous succession of horrors which confusedly succeeded each other for apparent eternities of being, and ended with some nameless catastrophe of woe or wickedness, in a waking more fearful than the state volcanically ruptured by it. During the nights I sat by him these occasional relaxations, as I learned, reached their maximum length, my familiar presence acting as a sedative, but from each of them he woke bathed in perspiration from sole to crown ; shivering under alternate flushes of chill and fever ; mentally confused to a degree which for half an hour rendered every object in the room unnatural and terrible to him ; with a nervous jerk, which threw him quite out of bed, although in his waking state two men were requisite to move him ; and with a cry of agony as loud as any under amputation.

The result of our consultation was a unanimous agreement not to press the case further. Physicians have no business to consider the speculative question, whether death without opium is preferable to life with it. They are called to keep people on the earth. We were convinced that to deprive the patient longer of opium would be to kill him. This we had no right to do without his consent. He did not consent, and I gave him five grains of morphia * between 8

* To the younger men of the profession rather than to the public gener-

and 12 o'clock on the morning of the day I had to return here. He was obliged to eat a few mouthfuls of sago before the alkaloid could act upon his nervous system. I need only point out the significance of this indication. The shallower-lying nervous fibres of the stomach had become definitely paralyzed, and such *digestion* as could be perfected under these circumstances was the only method of getting the stimulant in contact with any excitable nerve-substance. In other words, mere absorbent and assimulative tissue was all of him which for the purpose of receiving opium partially survived disorganization of the superficial nerves. Of that surviving tissue, one mucous patch was irredeemably gone. (This particular fact was the one which cessation from opium more distinctly unmasked.) At noon he had become tolerably comfortable ; before I left (7 P.M.) he had enjoyed a single half-hour of something like normal slumber.

He will have to take opium all his life. Further struggle is suicide. Death will probably occur at any rate not from an attack of what we usually consider disease, but from the disintegrating effects on tissue of the habit itself. So, whatever he may do, his organs march to death. He will have to continue the habit which kills him only because abandoning it kills him sooner ; for self-murder has dropped out of the purview of the moral faculties and become a mere animal question of time. The only way left him to preserve his intellectual faculties intact is to keep his future daily dose at the tolerable minimum. Henceforth all his dreams of entire liberty must be relegated to the world to come. He may be

ally I need here to say that this dose is not as excessive as it would naturally appear to be in the case of a man who had used no form of opium for ninety days. When you have to resume the drug, go cautiously. But you will generally find the amount of it required to produce the sedative effects in any case which returns to opium, after abandonment of a long habituation, *startlingly large,* and *slow in its effects.*

valuable as a monitor, but in the executive uses of this mighty modern world henceforth he can never share. Could the immortal soul find itself in a more inextricable, a more *grisly* complication ?

In publishing his case I am not violating that Hippocratic vow which protects the relations of patient and adviser ; for, as I dropped my friend's wasted hand and stepped to the threshold, he repeated a request he had often made to me, saying :

" It is almost like Dives asking for a messenger to his brethren ; but tell them, tell *all young men,* what it is, ' that they come not into this torment.' "

Already perhaps—by the mere statement of the case—I might be considered to have fulfilled my promise. But since monition often consists as much in enlightenment as intimidation, let me be pardoned for briefly presenting a few considerations regarding the action of opium upon the human system while living, and the peculiar methods by which the drug encompasses its death.

WHAT IS OPIUM?

It is the most complicated drug in the Pharmacopœia. Though apparently a simple gummy paste, it possesses a constitution which analysis reveals to contain no less than 25 elements, each one of them a compound by itself, and many of them among the most complex compounds known to modern chemistry. Let me concisely mention these by classes.

First, at least three earthy salts—the sulphates of lime, alumina, and potassa. Second, two organic and one simpler acid—acetic (absolute vinegar), meconic (one of the most powerful irritants which can be applied to the intestines through the bile), and sulphuric. All these exist uncombined in the gum, and free to work their will on the mucous tissues.

A green extractive matter, which comes in all vegetal bodies developed under sunlight, next deserves a place by itself, because it is one of the few organic bodies of which no rational analysis has ever been pretended. Though we can not state the constitution of this chlorophyl, we know that, except by turning acid in the stomach, it remains inert on the human system, as one might imagine would happen if he swallowed a bunch of green grass. *Lignin,* with which it is always associated, is mere woody fibre, and has no direct physical action. In no instance has any stomach been found to *digest* it save an insect's—some naturalists thinking that certain beetles make their horny wing-cases of that. I believe one man did think he had discovered a solvent for it in the gastric juice of the beaver, but that view is not widely entertained. So far as it exists in opium it can only act as a foreign substance and a mechanical irritant to the human bowels. Next come two inert, indigestible, and very similar gummy bodies, *mucilagin and bassorine.* Sugar, a powerfully active volatile principle, and a fixed oil (probably allied to turpentine) are the only other invariable constituents of opium belonging to the great organic group of the hydro-carbons.

I now come to a group by far the most important of all. Almost without exception the vegetable poisons belong to what are called the "nitrogenous alkaloids." Strychnia, brucia, ignatia, calabarin, woovarin, atropin, digitalin, and many others, including all whose effect is most tremendous upon the human system, are in this group. Not without insight did the early discoverers call nitrogen *azote,* "the foe to life." It so habitually exists in the things our body finds most deadly that the tests for it are always the first which occur to a chemist in the presence of any new organic poison. The nitrogenous alkaloids owe the first part of their name to the fact of containing this element; the second part to that of their usually making neutral salts with acids, like an alkaline base. The general reader may sometimes have asked himself why these alkaloids are diversely written—as, e. g., sometimes " *morphia,*" and sometimes " *morphine.*" The chemists who regard them as alkalies write them in the one way, those who consider them neutrals, in the other. Of these nitrogenous alkaloids, even the nuts of the tree, which furnishes the most powerful, *swift* poison of the world, contains but three— the above-named strychnia, brucia, and ignatia—principles shared in common with its pathological congener, the St. Ignatius bean. Opium may be found to contain *twelve* of them; but as one of these (cotarnin) may be a product of distillation, and the other (pseudo-morphia) seems only an occasional constituent, I treat them as ten in number—rationally to be arranged under three heads.

First, those whose action is merely acrid—so far as known expending themselves upon the mucous coats. (*Pseudo-morphia* when it occurs belongs to these.) So do *porphyroxin ; narcein ;* probably *papaverin* also ; while *meconin,* whose acrid properties in contact with animal tissue are similar to that of meconic acid, forms the last of the group.

The second head comprises but a single alkaloid, variously called paramorphia or thebain. (It may interest amateur chemists to know that its difference from strychnia consists only in having two less equivalents of hydrogen and six of carbon—especially when they know how closely its physical effects follow its atomic constitution.) A dose of one grain has produced tetanic spasms. Its chief action appears to be upon the spinal nerves, and there is reason to suppose it a poison of the same kind as nux vomica without the concentration of that agent. How singular it seems to find a poison of this totally distinct class—bad enough to set up the reputation of any one drug by itself—in company with the remaining principles whose effect we usually associate with opium and see clearest in the ruin of its victim !

The remainder, five in number, are the opium alkaloids, which act generally upon the whole system, but particularly, in their immediate phenomena, upon the brain. I mention them in the ascending order of their nervine power : narcotin ; codein ; opianin ; metamorphia, and morphia.

The first of these the poppy shares in common with many other narcotic plants—tobacco the most conspicuous among the number. In its anti-periodic effects on the human system it has been found similar to quinia, and it is an undoubted narcotic poison acting on the nerves of organic life, though, compared with its associates in the drug, comparatively innocent.

The remaining four act very much like morphia, differing only in the size of the dose in which they prove efficient. Most perfectly fresh constitutions feel a grain of morphia powerfully ; metamorphia is soporific in half-grain doses ;* opianin in its physical effects closely approximates morphia ; codein is about one-fifth as powerful ; a new subject may not get sleep short of six grains ; its main action is expended on the sympathetic system. It does not seem to congest the brain as morphia does ; but its action on the biliary system is probably little less deadly than that of the more powerful narcotic.

Looking at the marvellous complexity of opium we might be led to the *a priori* supposition that its versatility of action on the human system must be equally marvellous.

Miserably for the opium-eater, fortunately for the young person who may be dissuaded from following in his footsteps, we are left in no doubt of this matter by the conclusions of experience. In practical action opium affects as

* American Journal of Pharmacy, September, 1861.

large an area of nervous surface, attacks it with as much intensity, and changes it in as many ways as its complexity would lead us to expect. I have pointed out the existence in opium of a convulsive poison congeneric with brucia. The other chief active alkaloids, five in number, are those which specially possess the cumulative property. Poisons of the strychnia and hydro-cyanic acid classes (including this just mentioned opium alkaloid, thebain) are swifter agents ; but this perilous opium quintette sings to every sense a lulling song from which it may not awake for years, but wakes a slave. Every day that a man uses opium these cumulative alkaloids get a subtler hold on him. Even a physician addicted to the practice has no conception how their influence piles up.

At length some terrible dawn rouses him out of a bad sleep into a worse consciousness. Though the most untechnical man, he must already know the disorder which has taken place in his moral nature and his will. For a knowledge of his physical condition he must resort to his medical man, and what, when the case is ten years old, must a practitioner tell the patient in any average case ?

" Sir, the chances are entirely against you, and the possession of a powerfully enduring constitution, if you have it, forms a decided offset in your favor."

He then makes a thorough examination of him by ear, touch, conversation. If enough constitution responds to the call, he advises an immediate entrance upon the hard road of abnegation.

If the practitioner finds the case hopeless he must tell the patient so, in something like these words :

" You have either suffered a disorganization of irreproducible membranes, or you have deposited so much improper material in your tissue that your life is not consistent with the protracted pain of removing it.

" One by one you have paralyzed all the excretory functions of the body. Opium, aiming at all those functions for their death, first attacked the kidneys, and with your experimental doses you experienced a slight access of *dysouria*. As you went on, the same action, progressively paralytic to organic life, involved the liver. Flatulence, distress at the epigastrium, irregularity of bowels, indicated a spasmodic performance of the liver's work which showed it to be under high nervous excitement. Your mouth became dry through a cessation of the salivary discharge. Your lachrymal duct was parched, and your eye grew to have an *arid* look in addition to the dullness produced by opiate contraction of the pupil.

" All this time you continued to absorb an agent which directly acts for what by a paradox may be called fatal conservation of the tissues. Whether through its complexly combined nitrogen, carbon, or both, the drug has interposed itself between your very personal substance and those oxidations by which alone its life can be maintained. It has slowed the fires of your whole system. It has not only interposed but in part it has substituted itself; so that along with much effete matter of the body stored away there always exists a certain undecomposed quantity of the agent which sustains this morbid conservation.*

" When this combination became established, you began losing your appetite because no substitution of fresh matter was required by your body for tissue wrongly conserved. The progressive derangement of your liver manifested itself

---

* I frequently use what hydropaths call " a pack " to relieve opium distress, and with great benefit. After an hour and a half of perspiration, the patient being taken out of his swaddlings, I have found in the water which was used to wash out his sheet enough opium to have intoxicated a fresh subject. This patient had not used opium for a fortnight.

in increased sallowness of face and cornea ; the organ was working on an inadequate vital supply because the organic nervous system was becoming paralyzed ; the veins were not strained of that which is the bowels' proper purgative and the blood's dire poison. You had sealed up all but a single excretory passage—the pores of the skin. Perhaps when you had opium first given you you were told that its intent was the promotion of perspiration but did not know the *rationale.* The only way in which opium promotes perspiration is by shutting up all the other excretory processes of the body, and throwing the entire labor of that function upon the pores. (When the skin gives out the opium-eater is shut up like an entirely choked chimney, and often dies in delirium of blood-poisoning.)

"For a while—the first six years, perhaps—your skin sustained the work which should have been shared by the other organs—not in natural sweat, but violent perspiration, which showed the excess of its action. Then your palms became gradually hornier—your whole body yellower—at the same time that your muscular system grew tremulous through progressively failing nervous supply.

"About this time you may have had some temporary gastric disturbance, accompanied with indescribable distress, loathing at food, and nausea. This indicated that the mucous lining of the stomach had been partially removed by the corrosions of the drug, or that nervous power had suddenly come to a stand-still, which demanded an increase of stimulus.

"Since that time you have been taking your daily dose only to preserve the *status in quo.* The condition both of your nervous system and your stomach indicate that you must always take some anodyne to avoid torture, and *your* only anodyne is opium.

" The rest of your life must be spent in keeping comfortable, not in being happy."

Opium-eaters enjoy a strange immunity from other disease. They are not liable to be attacked by miasma in malarious countries ; epidemics or contagions where they exist. They almost always survive to die of their opium itself. And an opium death is usually in one of these two manners :

The opium-eater either dies in collapse through nervous exhaustion (with the blood-poisoning and delirium above-mentioned), sometimes after an overdose, but oftener seeming to occur spontaneously, or in the midst of physical or mental agony as great and irrelievable as men suffer in hopeful abandonment of the drug, and with a colliquative diarrhea, by which—in a continual fiery, acrid discharge—the system relieves itself during a final fortnight of the effete matters which have been accumulating for years.

Either of these ends is terrible enough. Let us draw a curtain over their details.

Opium is a corrosion and paralysis of all the noblest forms of life. The man who voluntarily addicts himself to it would commit in cutting his throat a suicide only swifter and less ignoble. The habit is gaining fearful ground among our professional men, the operatives in our mills, our weary sewing-women, our fagged clerks, our disappointed wives, our former liquor-drunkards, our very day-laborers, who a generation ago took gin. All our classes from the highest to the lowest are yearly increasing their consumption of the drug. The terrible demands especially in this country made on modern brains by our feverish competitive life, constitute hourly temptations to some form of the sweet, deadly sedative. Many a professional man of my acquaintance who twenty years ago was content with his *tri-diurnal* " whisky," ten years ago, drop by drop, began taking stronger " laudanum cock-tails," until he became what he is now—an habitual opium-eater. I have tried to show what he will be. If this article shall deter any from an imitation of his example or excite an interest in the question—" *What he shall do to be saved ?*"—I am content.

NOTE.—The patient whose sorrowful case suggested this article died just as the magazine was issued. His unassisted struggle had been too long protracted after abandonment of the drug was evidently hopeless, and his resumption of opium came too late to permit of his rallying from his exhaustion.

No. 1 Livingston Place,
Stuyvesant Square,
April 25, 1868.

MY DEAR SIR :—In accordance with your request, I sketch the brief outline of my plan for the treatment of opium-eaters, premising that it pretends much less to novelty than to such value as belongs to generalizations made from large experience by sincere interest and careful study in the light of science and common sense.

That experience having shown me how impracticable in the large majority of cases is any cure of a long-established opium habit while the patient continues his daily avocations and remains at home,* I shall simplify my sketch by supposing that one great object of my life is already attained, and that an institution for the treatment of the disease is already in successful operation. Starting at this fictitious *datum*, I shall carry from his arrival under our care until his

---

* In my article upon opium-eating, entitled, " What Shall They Do to be Saved ?" published in *Harper's Magazine* for the month of August, 1867, and hereto prefixed, I have referred to this impracticability in fuller detail. It arises from the fact that in his own house a man can not isolate himself from the hourly hearing of matters for which he feels responsible, yet to which he can give no adequate attention without his accustomed stimulus ; that his best friends are apt to upbraid him for a weakness which is not crime but disease, and that the control of him by those whom he has habitually directed, however well-judged, seems always an harassment.

discharge a healthy, happy, and useful member of society, a gentleman whom for convenience we will name Mr. Edgerton.

Our institution is called not an "Asylum," nor a " Retreat," nor by any of those names which savor of restraint and espionage—not even a " Home," as spelled with a capital H— but simply by the name of the spot upon which it is erected —to wit, " Lord's Island."

It is erected on an island because in the more serious cases a certain degree of watchfulness will always be necessary. On the main-land this watchfulness must be exercised by attendants with the aid of fences, bolts, and bars. On an island the patient whose case has gone beyond self-control will be under the Divine Vigilance, with more or less miles of deep water as the barrier between him and the poison by which he is imperilled. For this reason, and because whatever good is accomplished on it for a class which beyond all other sufferers claim heavenly mercy will be directly of the Lord himself, our island is called " Lord's Island." Here our patient will feel none of the irksome tutelage which in an asylum meets him at every step—thrusting itself before his eyes beyond any power of repulsion, and challenging him to efforts for its evasion which are noxious whether they succeed or not ; defeating the purpose of his salvation when they do, irritating him when they do not, and keeping his mind in a state of perpetual morbid concentration upon his exceptional condition among mankind in either case. Here he has all the liberty which is enjoyed by the doctors and nurses—save that he can not get at the medicine-chest.

Mr. Edgerton arrives at Lord's Island at 2 P.M. of a summer's day, having crossed by our half-hourly sail-boat, row-boat, or tug, from the railroad station on the main-land. If he is very much debilitated, either by his disease or fatigue, he has full opportunity to rest and refresh himself before a word is spoken to him professionally. If a friend accompanies him, he is invited to remain until Mr. Edgerton feels himself thoroughly at home in his new quarters.

After becoming fully rested, Mr. Edgerton is invited to state his case. The head physician must be particular to assure him that every word he utters will be regarded as in the solemnest professional confidence. Mr. Edgerton is made to feel that no syllable of his disclosures will ever be repeated, under any circumstances, even to the most intimate of his friends or the most nearly related of his family. This conviction upon his part is in the highest degree essential. Opium makes the best memory treacherous, and, sad as it may be to confess it, the most truthful nature, in matters relating to the habit at least, untrustworthy. Often, I am satisfied, the opium-eater, during periods of protracted effort or great excitement, takes doses of the drug which he does not recollect an hour afterward, and may, practically without knowing it, overrun his supposed weekly dose twenty-five per cent. I often meet persons addicted to the habit who, I have every reason to believe, honestly think they are using twelve grains of morphia daily, yet are found on close watching to take eighteen or twenty. Again, the opium-eater who by nature would scorn a lie as profoundly as the boy Washington, is sometimes so thoroughly changed by his habit that the truth seems a matter of the most trifling consequence to him, and his assertion upon any subject whatever becomes quite valueless. Occasionally this arises from an entire *bouleversement* of the veracious sense—similar to certain perversions of the insane mind, and then other faculties of his nature are liable to share in the alteration. If the man was previously to the highest degree merciful and sympathizing, he may become stolid to human suffering as any infant who laughs at its

N

mother's funeral, not from wickedness of disposition but absence of the faculty which appreciates woe, and I doubt not
that this change goes far to explain the ghastly unfeelingness
of many a Turkish and Chinese despot whose ingeniously
cruel tortures we shudder to read of scarcely more than the
placidity with which he sees them inflicted. If he was originally so sensitive to the boundaries between Meum and
Tuum that the least invasion of another's property hurt him
more than any loss of his own, this delicate sense may become blunted until he commits larceny as shamelessly as a
goat would browse through a gardener's pickets, or a child
of two years old help himself to a neighbor's sugar-plums.
This, too, quite innocently, and with the excuse of as' true a
Kleptomania as was ever established in the records of medical jurisprudence. I knew a man who had denied himself
all but the bare necessaries of life to discharge debts into
which another's fraud had plunged him, and whose sense of
honor was so keen that when afflicted with chronic dyspepsia
the morbid conscientiousness which is not an unusual mental
symptom of that malady took the form of hunting up the
owner of every pin he picked up from the floor, nor could he
shake off a sense of criminality till he had found somebody
who had lost one and restored it to him—yet on being prescribed opium for his complaint, his nature, under its operation, suffered such an entire inversion that the libraries, and
on several occasions even the pocket-books of his friends
were not safe from him, his larcenies comprising some of
the most valuable volumes on the shelf and sums varying
between two and twenty dollars in the porte-monnaie. "The
Book-Hunter" writing of De Quincey, as you will recollect,
under the *sobriquet* of "Papaverius," describes the perfectly
child-like absence of all proprietary distinctions which prevailed in that wonderful man's mind during his later years as

regarded the books of his acquaintance, and the innocent way in which he abstracted any volume which he wanted or tore out and carried away with him the particular leaves he wished for reference.

In many cases where the moral sense has suffered no such general *bouleversement*, the tendency which opium superinduces to look at every thing from the most sanguine point of view —the vague, dreamy habit of thought and the inability to deal with hard facts or fixed quantities—make it necessary to take an opium-eater's assertions upon any subject with a certain degree of allowance—to translate them, as it were, into the accurate expressions of literal life ; but even where this necessity does not exist, in cases sometimes though rarely met with, where opium has been long used without tinging any of life's common facts with uncertainty, an opium-eater can scarcely even be relied on for the exact truth concerning his own habit. He may be trusted without hesitation upon every other subject, but on this he almost always speaks evasively, and though about any thing else he would cut his hand off rather than say the thing that is not, will sometimes tell a downright falsehood. In most cases he has been led to this course by witnessing the agony or suffering the reproach with which the knowledge of his habit is received by his friends. He lies either in mercy to them or because the pangs which their rebuke inflicts would become still more intolerable if they knew the extent of his error.

It is therefore always proper that the opium-eater should find in his physician a confidant who will not violate his secret even to parent or wife. The closer the relation and the dearer the love, the greater will be the likelihood that the opium-eater has shrunk from revealing the full extent of his burden to the friend in question, and the greater will be the temptation to deceive the doctor unless the patient be made to feel that his revelation is as sacred as the secrets of the bridal-chamber.

I solicit from the friend who accompanied Mr. Edgerton the thoroughest statement which he can give me of the case, *ab extra.* Such a statement is of great value—for the inroads which the habit has made upon the system are often visible to an outsider only. Furthermore, a friend may give me many circumstances connected with the inception of the case : family predispositions and inherited tendencies ; causes contributing to the formation of the habit, such as domestic or business misfortune, prior bad habits of other kinds, illnesses suffered, and a variety of other agencies concerning which the patient might hesitate or forget to speak for himself. Then I make Mr. Edgerton the proffer of that inviolable confidence which I have mentioned, and having won his perfect faith in me, obtain the very fullest history of his case which can be elicited by searching, but most kindly and sympathizing cross-examination. The two statements I collate and enter for my future guidance in a private record.

Let us suppose an average hopeful case.

I find that my patient is about thirty years of age—of the energetic yet at the same time delicate and sensitive nervous organization which is peculiarly susceptible to the effects of opium, from which it draws the vast majority of its victims, and in which it makes its most relentless havoc ; with a front brain considerably beyond the average in size and development. My patient's general health, apart from the inevitable disturbances of the drug, has always been fair, and his constitution, under the same limitations, is a vigorous one. His habit, as in nine cases out of every ten, dates from the medical prescription of opium for the relief of violent pain or the cure of obstinate illness. He was not aware of the drug then administered to him, or at any rate of the peril attending its use, and his malady was so long protracted that opium had established itself as a necessary condition of comfortable existence before he realized that it possessed the slightest hold upon him. When the prescription was discontinued he suffered so much distress that he voluntarily resumed it, without consulting his physician, or, if he did consult him, receiving no further warning as to his danger than that "he had better leave off as soon as practicable." Or else, on leaving off his use of opium, the symptoms for which it had originally been administered returned with more or less severity, and under the idea that they indicated a relapse instead of being one of the characteristic actions of the drug itself, he resumed the dose. It gradually lost its power; little by little he was compelled to increase it; and having begun with ⅓ grain powders of which he took three per diem, he is now taking 18 grains of morphia per diem at the end of five years from his first dose.

If I find him tolerably vigorous on his arrival, as will be the case when he has come to Lord's Island after calm deliberation and the conviction not that he *must*, but on all accounts *had better* abandon the habit, I leave him to recover from the fatigues of his journey and get acquainted with his surroundings before I begin any treatment of his case. If, however, as sometimes occurs, he reaches us in desperate plight, having been so far injured by his habit as to show unequivocal signs of an opium-poisoning which threatens fatal results; if, as in several cases known to me, he has summoned all his remaining vitality to get to a place of refuge, being overtaken either by that terrible *coma* which often terminates the case of the opium-eater in the same fashion that persons new to the narcotic are killed by an overdose, or by that only less terrible opium-delirium belonging to the same general class as mania potu—then his case admits of not a

moment's delay.   Opium-eaters differ so widely—every new
case furnishing some marked idiosyncrasy which may demand an entirely different management and list of remedies
from those required by the last one—that for any general
scheme of treatment a week's study of the patient will be
necessary.   During that week our attitude will be simply tentative and expectant, and at its close the proper fidelity and
vigilance will have authorized us in making out something
like a permanent schedule for the patient's upward march,
though even then we must be prepared, like skillful generals,
to meet new emergencies, take unforeseen steps, even throw
overboard old theories, at any stage of his progress.   In no
disease is there such infinite variety as in that of opio-mania,
in none must the interrogation of nature be more humbly def-
erent and faithfully attentive; in none do slight differences of
temperament, previous habits, and circumstances necessitate
such wide variation in the remedies to be used.   Notice, by
way of illustration, the fact that one opium-eater under my
care was powerfully affected and greatly benefited by the
prescription of *one drachm* of the fluid extract of *cannabis
indica*, while another, in temperament, history, tendencies, and
all but a few apparently trifling particulars almost identical,
not only received no benefit but actually experienced no perceptible effect whatever from the absolutely colossal dose of
*four fluid ounces.**   I may add that in the latter case, *bro-*

---

* I am aware how incredible this statement will seem to those who
have never had any extensive experience of the behavior of this remarkably variable drug, and get their notion of its action from the absurd directions on the label of every pound vial I have seen sent forth by our
manufacturing pharmaceutists.   "Ten to twenty drops at a dose," they
say, "cautiously increased."   Cannabis should always be used with caution, but ten or even twenty drops must be inert in all but the rarest
cases, and I have given an ounce per diem with beneficial effect.   But
four ounces of the best extract (Hance & Griffith's) producing literally

mide of potassium was administered with the happiest result—
in fact as nearly approaching in its efficiency the character
of a succedaneum as any remedy I ever used to alleviate the
tortures of opium, while in the former no result attended its
administration salutary or otherwise. The vast diversity of
operation exhibited in different patients by the drug *scutellaria* is still another illustration of the careful study of idiosyncrasies requisite for a successful treatment of the opium
disease. But when the case comes into our hands at a desperate period there are many means of instant alleviation
which may anticipate without interfering with future treatment based on study.

Mr. Edgerton, though by no means a man of ruined constitution, has brought himself temporarily into a critical place
by the fatigues and anxieties of harassing business, by exceptional overwork which kept him at his desk or in his
shop until inordinately late hours ; even, let me say, by going
for entire nights without sleep and neglecting his regular
meals day after day for a period of several weeks ; performing and enduring all this by the support of extra doses of
opium. Perhaps, finding the stimulus to which he has become accustomed too slow in its operation, he has violated
his usual custom of abstinence from alcoholic drinks and reinforced his opium with more or less frequent potations of
whisky. This is no fancy sketch. Our overtasked com-
mercial men frequently go on what might with propriety be
called "a business spree," in which for a month at a time,
whether using stimulants or not, they plunge into as mad a
vortex with as thorough a recklessness as those of the periodical inebriate ; finding out in the long run that the fascina-

no effect of any kind on an entirely fresh subject, is a phenomenon that I
must have needed eye-witness to imagine possible.

tions of speculation, and the spring and fall trade, bring as dire destruction to soul and body as those of the bowl and the laudanum vial. During times of great financial pressure or under the screws of preparation for some great professional effort, the moderate opium-eater finds that he must inevitably increase his dose. When he adds liquor to it (and this addition to an old opium-eater is often as necessary as liquor alone would have been before he used opium at all) he is indeed burning his candle at both ends. Mr. Edgerton reached the commencement of his period of extra exertion with as sound a constitution—in as comfortable condition of general health—as is enjoyed by any man habituated to opium for four or five years ; and such cases are frequently found among men who appear to enjoy life pretty well, attend to their business with as much regularity as ever, and show no trace of the ravages wrought by their insidious foe to any but the expert student. After six weeks of exciting labor and solicitude, during which his sleep and his rations were always delayed till exhaustion overpowered him, and then cut down below half the normal standard, he wakes one morning from a slumber heavy as death into a state of the most awful vigilance his mind can conceive of. He even doubts for some moments whether he shall ever sleep again, and in the agony of that strange, wild suspicion, a cold sweat breaks out over him from head to foot. Waking from the most utter unconsciousness possible to a wide-awake state like having the top of one's skull suddenly lifted off by some surgeon Asmodeus, and the noonday sun poured into every cranny of his brain, he suffers a shock compared with which any galvanic battery, not fatal, gives but a gentle tap. The suddenness of the transition—no gentle fading out of half-remembered dreams, no slow lifting of lids, no pleasant uncertainty of time and place gradually replacing

itself by dawning outlines of familiar chair and window frame
and cornice—the leap from absolute nonentity into a glar-
ing, staring world—for a moment almost unsettles Mr. Edgerton's reason. Then the fear for his sanity passes and a
strange horror of approaching death takes its room. His
pulse at the instant of waking throbs like a trip-hammer ;
an instant more and it intermits. Then it begins again at
the old pace. He snatches up his watch from the bureau with
a trembling hand and counts—the beat is 130 a minute.
Again it stops ; again it begins ; but now little by little growing faster and threadier until it runs so swiftly yet so thinly
as to feel under his finger like some continuous strand of
gossamer drawn through the artery. His feet and hands
grow deadly cold. He seems to feel his blood trickling feebly back to his heart from every portion of his body. He
catches a hurried look at the glass—he sees a dreadful spec-
tre with bistre rings around the eyelids, an ashen face, leaden
lips, and great, mournful, hollow, desolate eyes. Then his
pulse stops altogether ; his lungs cease their involuntary
action ; and with a sense of inconceivable terror paralyzing
the very effort he now feels it vital to make, he puts them
under voluntary control and makes each separate inspiration
by an effort as conscious as working a bellows. I doubt not
that many men have died just at this place through absolute
lack of will to continue such effort. Then the metaphorical ·
paralysis of fear is seconded by the simulation of a literal
one, extending through the limbs of one side or both ; the
sufferer reels, feeling one foot fail him—tries to revolve one
arm like a windmill, that he may restore his circulation, and
that arm for some instants hangs powerless. Presently, with
one tremendous concentration of will, his brain shouts down
an order to the rebellious member—it stirs with sullen reluctance—it moves an inch—and then it breaks from the

prison of its waking nightmare. Summoning his entire array of vital forces, our patient leaps, and smites his breast, kicks, whirls his arms, and little by little feels his heart tick again. By the time a feeble and sickly but regular pulse is re-established he has gone through enough agony to punish the worst enemy, my dear Sir, that you or I ever had. The vague, overpowering fear of death which during such an attack afflicts even the man who by grace or nature is at all other times most exempt from it is one of this period's most terrible symptoms. This passes with the return of breath and circulation.

But the clammy sweat continues—pouring from every point of the surface—saturating the garments next the skin as if they had been dipped in a tub of water. Presently our patient begins to suffer an intolerable thirst, and runs to the ice-pitcher to quench it. In vain. He can not retain a mouthful. The instant it is swallowed it seems to strike a trap and is rejected with one jerk. He seeks the sedative which up to this hour has allayed his worst gastric irritations. Now, if never before, opium in every form produces nausea. Laudanum instantly follows the example of the water, and even a dry dose of morphia, swallowed with no moisture but saliva, casts itself back after agonizing retchings. To liquor his rebellious stomach proves yet more intolerant —food is almost as irritating as liquor. In a horror he discovers that even pounded ice will not stay down—and he is parching like Dives. His anguish becomes nearly suicidal as the fact stares him in the face that he has come to the place where he can not take opium any more—though to be without it is hell—that food, drink, medicine, are all denied him.

A merciful, death-like apathy ensues. He lies down, and with his brain full of delirious visions, appalling, grotesque, meaningless, beautiful, torturing by turns, still manages to catch an occasional minute of unconsciousness. He hears his name called—tries to rise and answer—but his voice faints in his throat and he falls back upon his bed. Friends enter his bed-chamber—in an agony of alarm rouse him—lift him to his feet—but he has not the strength of an infant, and he falls again. In this condition he may continue for a day or two, then sink into absolute coma, and die of nervous exhaustion, or his constitution may rally as the effects of the last overdose pass off, and the man, after a fortnight's utter prostration, come gradually back to such a state of tolerable health and comfort as he enjoyed before he overtaxed himself.

Mr. Edgerton is brought to Lord's Island in the condition I have described, living near enough to be transported on mattresses in carriage and boat. A few hurried questions put to his friends reveal that although his condition is alarming it is by no means necessarily fatal; being one of those in which the habit is of such comparatively short standing, and the constitution still so vigorous, that even at home he might come up again by natural reactions.

He is immediately undressed and put to bed, with hot bricks and blankets at the extremities, and the galvanic battery is judiciously administered by placing both feet in contact with a copper plate constituting the negative electrode, while the operator grasps the positive in one hand, and having wetted the fingers of the other, follows the spine downward, exerting gentle pressure with them as he goes. "Judiciously," I say, because there is a vast deal of injudicious use of the battery. In many cases, for instance, a powerful and spasmodic current is used to the absolute injury of the patient, where the greatest benefit might be secured by an even one so light as scarcely to be perceptible. But I can only mention the battery. Its application is by itself a science, and demands a book. The practitioner who treats opium patients needs that science as much as any one interested in whatsoever branch of nervous therapeutics. The battery in the hands of a scientific man is one of our most powerful adjuncts throughout every stage of treatment, both of opium-eating and its sequelae. Paralysis following the habit, and persistent long after its abandonment, I have cured by it when all other means failed. Here, however, we have only room to indicate the weapons in our armory.

If Mr. Edgerton's digestive apparatus is still as intolerant as at the commencement of the attack which hurried him to Lord's Island, we may hope for a marked mitigation of this symptom, in the use of the battery by passing a mild current transversely through him in the region of the solar plexus. As soon as it is possible for his stomach to retain any thing we administer a bolus of *capsicum*, compounded of five grains of the powder with any simple addition like mucilage and liquorice to make it a coherent mass. The remaining nausea and irritability will in great likelihood be speedily relieved as by magic, and with these will disappear some of the most distressing cerebral symptoms — the horror and frenzy or comatose apathy among them. In few cases will a patient reach the Island in time for the advantageous use of *belladonna*. That is a direct antidote—exerting its function in antagonism to the earlier toxical effects of the opium. In cases where a single overdose has worked the difficulty and produced the coma which Mr. Edgerton's now resembles, it may be given to an old *habitué* of the drug with as good advantage as to a person whose overdose is his first experience of opium. It is of especial value where the absorbents have carried the excess beyond the reach of an emetic, any time, indeed, within fifteen or twenty hours after the overdose, when sulphate of zinc and the stomach-pump have failed to bring the poison. If our patient on the Island has taken his overdose so recently, and it seems still worth while to act by antidote, we shall be obliged to get over the difficulty presented by his stomach's lack of retention by administering our belladonna in the form of *atropin* in solution as a hypodermic injection. The many eminent researches of late made in this interesting method of administering remedies, and the practitioner's own judgment, must guide him as to the proportions of his dose—whether one-fortieth grain, one-twentieth, or larger. Of this operation, with opium-eaters, I have seen several most successful instances.

In all probability, however, there will be a better field in such cases as Mr. Edgerton's for the use of nux vomica than of belladonna. Where the prostration is so great as to call for the most immediate action to avoid a syncope from which there shall be no rallying, it will be unwise to await the soothing action of the battery, capsicum, or any other means preparatory to giving nux vomica by the mouth. *Strychnia* in solution (it is needless to say with what caution) must be administered like the *atropin*, subcutaneously, or else nux vomica tincture in the form of the ordinary enema *per rectum* in about the same dose as it would be given by the mouth. The former method in wise hands is the better, both as the speedier, and, considering the opiate torpidity of the intestines, by far the more certain. In cases where the stomach tolerates fluid, as our ability to await the action of the battery and capsicum have now enabled us to find Mr. Edgerton's, we may give from fifteen to twenty drops of the ordinary pharmaceutical tincture of nux vomica in a table-spoonful of water.

In the course of ten minutes we find a decided improvement in the pulse of the patient ; he experiences great relief from his feelings of apprehension and distress about the epi-gastrium; and the most powerful tonic known to science begins dispatching its irresistible behests to every fibre of the organic life. That painful as well as agitating *subsultus* —that involuntary twitching and cramp in the muscles of the -limbs and abdomen which often characterizes this form of the opium malady, by degrees gets lulled as under a charm, and it may not even be necessary to repeat the dose in two and a half hours to remove it so entirely that the patient gets ten or fifteen minutes of refreshing sleep.

The earliest symptoms of this species of attack sometimes indicate such prostration as make any bath of the ordinary kind unsafe; yet rare indeed are the cases (not one in a hundred I should say) where there is any danger of further depressing the nervous system (of course the great thing to guard against) by putting a patient like Mr. Edgerton into a *Russian bath*. I need not enlarge upon the value of this most admirable appliance—all the most enlightened men of the medical profession know it and esteem it as it deserves, though its use in rheumatic affections and cutaneous diseases has hitherto received more study than in the class of mala-dies where its employment is perhaps the most beneficial of all—the nervous. Pre-eminently valuable is it in the treat-ment of delirium tremens and in every stage of the opio-mania.

As your book is for the purpose of the public rather than professional men, I may perhaps properly say a few words about this bath by way of description. We have one, as a matter of course, at Lord's Island.

A room forty-five feet long and twenty broad, with a vault-ed ceiling twenty feet high at the crown, is provided along each of its two longer sides with a series of marble slabs rising in three tiers from eighteen inches above the floor to a couple of feet below the ceiling. The idea may be gained more accurately by supposing three steps of a giant staircase mounting from an aisle three feet wide through the middle of the room, back and upward to the parallel cornice. The level surface of each of these steps is sufficiently wide to accommodate a man stretched on his back, and the upright portion of each step is an iron grating. Under the series of steps on both sides runs a system of sinuous iron pipes pierced with minute holes, and connected by stop-cocks with a boiler out of sight.

The steps occupy in length twenty-five feet of the room, and its entire breadth except the narrow aisle between the two series. The remaining twenty feet square of the room is occupied by a tank sunk beneath the floor, sixteen feet square by four and a half deep, filled with water kept throughout the year at a uniform temperature of about 70° F., and by the gallery which runs round the railing of the tank on the floor level. About the sides of the gallery are arranged hot and cold water-pipes with faucets and hose connections, the hose being terminated by a spray apparatus similar to the nose of a watering-pot. Opening off the gallery at the end furthest from the steps is a small closet fitted up with ascending, descending, and horizontal shower apparatus, by means of perforated plates connecting with the water-pipes by faucets set in floor, walls, and ceiling.

After the battery, the capsicum, and the nux, if Mr. Edgerton can retain it, we feed him by slow tea-spoonfuls from one-half to a whole cup of the most concentrated beef-tea—prepared after Liebig's recipe or another which I have usually found better relished, and as that, where food must be administered to fastidious stomachs, is half the battle, which I prefer. (I will give it hereafter.) Should his stomach reject it thus administered, it must be given as an enema. Its place in the plan of all enlightened medical treatment is too lofty to need my insisting on. We must rely on it at Lord's Island every step of our way. It will not have been within our patient's system five minutes before the pulse shows it, nor ten before he feels from head to foot as if he had taken some powerful and generous stimulant. It is always wise to give beef-tea, even just before a bath of any kind, and it is never well to enter the Russian bath on an empty stomach.

Having taken his beef-tea, Mr. Edgerton is carried or propelled in a wheel-chair by attendants to the Russian bathroom. Having stripped in an anteroom, upon entering the vaulted chamber he finds himself in an atmosphere of steam at 120° F., which fills the apartment, even obscures the skylights, yet to his surprise does not impede his respiration or produce any unpleasant sense of fullness in the head. He is now stretched on his back upon one of the lowest slabs, where the atmosphere is coolest and the vapor least dense ; a large wet sponge is put under his occiput for a pillow, and another sponge in a pail of cool water placed by his side with which he, or in case of too extreme debility his attendants, may from time to time bathe and cool the rest of his head. As soon as he has become accustomed to the heat and moisture, a sensation of pleasant languor steals over him ; all remains of his nausea and other gastric distress vanish ; his nervous system grows more and more placid ; his clammy skin is bedewed by a profuse and warm natural perspiration. Perhaps, as in cases of extreme debility and where the nerves have suffered tension from protracted pain, he even falls into a pleasant sleep. He is allowed to lie quietly on this lower slab for about fifteen minutes. An attendant then lathers him from head to foot with a perfumed cake of soap and gives him a gentle but thorough scrubbing with an oval brush like that in use among hostlers—finishing the operation by vigorously shampooing, Oriental fashion, each separate joint of his whole body, with a result of exquis- ite relief not exaggerated by Eastern travellers as applicable to well people and quite beyond expression when its subject is the poor, long-tortured frame of a sick opium-eater. This process over, the patient is taken to the gallery and stood up before the hose apparatus above-mentioned. One hand of the attendant directs over his body a fine spray of steam and the other follows it up and down with a spray of cool water (either of which by combining and graduating appropriate faucets may be made as warm as you like), producing a fine glow and reaction of the whole surface. The up, down, and lateral showers are then administered, after which the pa- tient is sent to plunge into the tank, and if able to swim, a stroke or two. Emerging, rosy as Aphrodité, and with a sense of vigor he can hardly believe, he again lies down on the slab—this time taking the next higher tier, and in about ten minutes more, mounting, if so disposed, to the highest, where the perspiration rolls from him in rivulets, and with it such an amount of poison seems discharged from his veins as makes him feel like a new being. Finally, in about an hour from the time he entered the bath-room he is treated to one last plunge in the tank and carried back to the anteroom. The thermometer there marks but 70° F., or half a hundred degrees cooler than the steam from which he has just emerged ; still his blood has been set in such healthful cir- culation, and during the last hour he has absorbed such an amount of caloric, that the change seems a very pleasant one, and his skin has been so toned that he runs not the slightest risk (even were he the frailest person with pulmonary dis- ease) of catching cold. Singular as it may seem, the first case of such a result has yet to be recorded.

This is all the more remarkable when we consider that instead of being immediately wrapped up after his vigorous drying with furzy bath-towels, he is kept naked for five minutes longer during a further process of hand-rubbing and shampooing by an attendant. The shampooing takes place as he lies prostrate on a couch and thus gives his debility all the advantage of rest and passive exercise at the same time. Whether we explain it upon the yet unsettled hypotheses of animal magnetism or upon simple principles of mechanical friction, the suppling which the patient gets in this part of the process from the hands of a strong, faithful, cheerful-minded and hale-bodied servant is one of the most valuable means which can be relied upon for the relief of opium suffering at any stage whatever. After coming from the ante-room our patient who entered more dead than alive may feel in possession of such a new capital of nervous and muscular vigor that he would like to give his recovered powers play in walking back to his room, but it is best not to humor him by letting him draw on his first deposit. He should be tenderly wheeled back as he came—put to bed, and if it does not revolt his appetite, fed slowly as before another cup of beef-tea. After that he will probably fall into a refreshing slumber from which he is on no account to be roused, but suffered to wake himself. On his waking another cup of beef-tea should be given him, and no other medicine, unless his pulse becomes alarming and he shows signs of return to the original sinking condition in which we found him—when the nux may be repeated.

It is now improbable, after the happy change described has taken place in him, that he will succumb to the acute attack of opium-poisoning which led him to us. Alarming as it appears, it is seldom dangerous or persistent. The patient who has not constitutional strength to rally at once, goes down rapidly and dies in a few days, while he who rallies once gets well, *pro hâc vice*, without much medical treatment save that which was promptly given at the critical moment, or treatment of any kind but nourishing food, rest, baths, and vigilant, tender nursing. As soon as the chronic appetite calls for its habitual dose, and the stomach receives it without revenging its grudge against the recent excesses, the patient may be considered out of danger as far as the acute attack is concerned.

Here I will be asked (as I am constantly out of the book), why not begin the abandonment of the drug as soon as this acute attack is over? When the terrible and immediate peril has been staved off by such a mere hair's-breadth, why listen again to "the chronic appetite" which "calls for its habitual dose?" Surely, now that the patient has gone for forty-eight hours or more without that dose, would it not be better never to return to it? Must he begin his former career again and afterward have all the same ground to go over?

I answer that he will not have the same ground. That which he has just traversed was the ground separating between an excess and his normal life—and he is in reality in a worse condition to try the experiment of instant abandonment than he was before the struggle. It is a very different thing to cure a man of acute from curing him of chronic opium-poisoning; and my own large experience, together with that of all the most experienced, the soundest and most skillful men that I have ever known as successful practitioners among these cases, points to the unanimous conclusion that it is not safe, either to mind or body, to make the abrupt transition required of an old opium-eater who must give up his drug *in toto* and at once, especially after such an acute attack as that just described. He would be very likely to die of exhaustion, to endure an amount of agony which would permanently enfeeble his mind, or to commit suicide as his only way of escape from it, if we cut him short from the equivalent of 15 or 20 grains of sulphate of morphia after having used the drug for five years. The most terrible case of opium-eating which I ever saw instantly cut off short was one where the patient used 32 grains of morphia per diem, but he had used it for less than a year, and possessed a constitution whose physical grit and mental pluck any body would pronounce exceptional, though even that did not save him from tortures which endangered his reason. I am always in favor of a man's "breaking off short" if he can. I believe that the majority of people who have used the drug less than a year can, but the number who are able to do it after that diminish in geometric ratio with every month of habituation.

I therefore permit Mr. Edgerton, as soon as his stomach will bear it, to return to the use of opium.

But before giving him his dose I make the stipulation that from this moment he shall deal as frankly with me as he does with his own consciousness — that we shall have no opium secrets apart.

In advanced cases, where opium has been used long enough to break down the will and the sense of moral accountability, I may feel it wise to ask of the friend who accompanies my patient that he shall go through the baggage and clothes of the latter before leaving him, and report to me that no form of opium is contained in them. But in most cases I prefer to rely entirely upon the good understanding established between my patient and myself for my guarantee that no opiate is smuggled into the institution, and upon my own daily examination of the patient to determine whether this guarantee is kept inviolate. To an expert reader of opium cases it will soon become apparent whether in any given case a patient is taking more than the amount prescribed—and after total abandonment is resolved upon, the question whether the patient is taking opium at all may be decided by a tyro.

In the case of Mr. Edgerton, who has voluntarily come to ask our help on the way upward, I proceed by a system of complete mutual confidence. I tell him that I am sure he feels even more deeply than myself the necessity of abandoning the drug. I promise him that he shall never be pushed beyond the limits of endurance, and ask only that he will allow any dose he may take to pass through my hands. I request that if he has brought any form of opium with him he will give it to me, and we enter into a stipulation that he will come to me for any opiate or other alleviative which he may desire. I bind myself never to upbraid or censure him—never to reveal to a living soul any confidence soever which he may repose in me—and then I ask him to name me the average dose upon which, before his late acute attack, he has managed to keep comfortable—rather, I should say, before the overwork and consequent opiate excess which brought it on. During his terrible six weeks of high-pressure, he tells me, he reached a per diem as high as 25, on one occasion even 30 grains ; but for a year previous he had never taken more than the equivalent of 18 grains of morphia a day. This, then, shall furnish our starting-point.

Whether he has previously adopted the same method or not, I divide this amount into three or more doses to be taken at regular intervals during the day.

I say " the equivalent of 18 grains of morphia," because although the majority of *habitués* use that principle of opium as their favorite form, there are some who after many years' use of the drug still adhere to crude gum opium or laudanum. The portability and ease of exhibition which belong to mor-

phia—the fact that it fails to sicken some persons in whom
any other opiate produces violent nausea—its usual certainty,
rapidity, and uniformity of action, and the ability which it
possesses to produce the characteristic effects of the narcotic
after other preparations have become comparatively inert,
make it the most general form in use among opium-eaters
of long standing.   Still, bearing in mind the wonderful complexity of opium (*vide* "What Shall They Do to be Saved?")
and the equally marvellous diversity in the manner in which
it affects different people, we can not wonder at the fact that
some of its victims require for their desired effect either the
crude drug or other preparations containing its principles
entire.   Morphia is by far the most important of these principles, and more nearly than any one stands typical of them
all.   Still, it is easy to conceive how certain constitutions
may respond more sympathetically to the complex agent of
Nature's compounding than to any one of its constituents.*
We may therefore find it necessary to carry on our reforma-
tory process upon laudanum or M'Munn's Elixir, but by far
the larger number of cases will do better by being put instantly upon a·regimen of Magendie's Solution of Morphia.
The formula for this preparation is :

℞
Morph. Sulph.   .   .   .   .   grs. xvi.
Aqua Destill.   .   .   .   .   ʒj
Elix. Vitrioli   .   .   .   .   quant. suff.

Mr. Edgerton has used 18 grains of morphia per diem.

* In some cases, especially of shorter standing, codeia may be used as
the form of opiate to diminish on.   In any case its employment is worth
trying, for it possesses much of the pain-controlling efficiency of opium
and morphia, with less of their congestive action upon the brain.   Practically it may be treated in such an experiment as the equivalent of opium ;
not that it at all represents all the drug's operations, but that where crude
opium has been the form in use, codeia may be substituted grain for grain.
Some patients find it quite valueless as a substitute, but there is always a

His equivalent in Magendie's Solution will be 9 fluid drachms.

.This amount I divide into three equal doses—one to be administered after each meal. By administering them after meals I give nutrition the start of narcotism, prevent the violent action possessed by stimulants and opiates on the naked stomach, and secure a slower, more uniform distribution of the effects throughout the day. The position of the third dose after the 6 o'clock meal of the day is particularly counselled by the fact that opium is only secondarily a narcotic, its sedative effects following as a reaction upon its stimulant, and the third dose accordingly begins to act soporifically just about bed-time, when this action is especially required.

I keep a glass for each of my patients, upon which their " high-water mark " is indicated by a slip of paper gummed on the outside. When Mr. Edgerton, pursuant to our stipulation, comes to me for his dose, I drop into the glass before his eyes a shot about the size of a small pea—then fill the glass with Magendie's Solution up to the mark indicated. (This shot varies in each case with the rapidity of diminution I think safe to adopt. In some cases it is a buckshot or a small pistol bullet.) Every day a new shot goes in—and if he bears that rate of progress I may even drop one into the glass with each alternate dose.

Midway between the doses of morphia I give Mr. E. a powder of bromide of potassium, amounting to 30 or even 40 grains at a time, and an average of about 100 grains per day. The value of this remedy has been a matter of much controversy—some practitioners lauding it to the skies as one of the most powerful agents of control in

chance of its proving adequate. When tried, the best form is a solution similar to Magendie's, but replacing one grain of morphia by six of codeia.

all disorders of the nervous system, others pronouncing it
entirely inert. Where it has proved the latter it has probably been given in too small doses or not persevered in for a
sufficient length of time. (The timidity with which it is oft-
en prescribed may be seen in the fact that one of the principal druggists on Broadway lately warned a person to whom
I had given a prescription for 30 grain doses that he was
running a very dangerous risk in taking such a quantity!)
Its operation is so entirely different from that of the vegetable narcotics that people looking for their instantaneous
sedative effect can not fail to be disappointed. It is very
slowly cumulative in its action, seeming to act upon the
nervous system by a gradual constitutional change rather
than any special impetus in a given direction. Because that
is its *modus operandi*, I begin to give it thus early; and it is
of peculiar value now, not only as making the daily diminution of the opium more tolerable, but as preparing the system
for the time when the drug is to be abandoned altogether
and the hardest part of the tug comes.

In Mr. Edgerton's case the gradual descent to ½ grain
per diem, when we leave off the opium entirely, consumes
let us say a period of one month. It is not to be expected that this period will pass without considerable discomfort and some absolute suffering, for the nervous system
can not be dealt with artfully enough to hide from it the fact
that it is losing its main support. It is the nature of that
system not even to rest content with the continuation of the
same dose. It grows daily less susceptible to opium and
more clamorous of increase. When the dose does not even
remain *in statu quo* but suffers steady diminutions however
small, the nerves can not fail to begin revenging themselves.
Still, this period may be made very tolerable by keeping the
mind diverted in every pleasant occupation possible, such as

I shall presently refer to as abounding on our Island. Our physical treatment for the month is especially directed to the establishment of such healthy nutrition and circulation as shall provide the nervous system with a liberal capital to draw on during the exhausting struggles which are to ensue for at least the first ten days or fortnight after the complete abandonment of opium. The patient's digestion must be carefully attended to, and kept as vigorous as is consistent with the still continued use of the drug. Beef-tea, lamb-broth with rice, all the more concentrated forms of nutriment, are to be given him, in small quantities at a time, as frequently as his appetite will permit ; and if progressive gastric irritability does not develop itself as the diminution of the narcotic proceeds, he is to have generous diet of all kinds. We must pay particular attention to the excretory functions— getting them as nearly as possible in complete working order for the extra task they have presently to fulfill when the barriers are entirely withdrawn and the long pent-up effete matters of the body come rushing forth at every channel. The bowels must be trained to perfect regularity, and the skin roused to the greatest activity of which it is capable. Exercise, carried to the extent of healthy fatigue, but rigorously kept short of exhaustion, may be secured in our bowling-alley, gymnasium, and that system of light gymnastics perfected by Dio Lewis—a system combining amusement with improvement to a remarkable degree, as being a regular drill in which at certain regular hours all those patients, both ladies and gentlemen, who are able to leave their rooms, join under the command of a skillful leader to the sound of music. This system has an advantage, even for well people, over the ordinary exercises of the old-fashioned gymnasium, with its bars, poles, ropes, dumb-bells, etc., inasmuch as it secures the uniform development, on sound anatomical and

O

physiological principles, of every muscle in the human body, instead of aiming at the hypertrophy of an isolated set. I do not mean by this to deny the value of the old style gymnasium, our Island will possess as good a one as any athlete could desire. Horseback riding will form another admirable means of effecting our purpose, especially where the patient suffers from more than the usual opiate torpidity of the liver. We shall have room enough if not for an extended ride at least for a mile track around the Island, and a stud, however unlikely to set John Hunter looking to his laurels, capable of affording choice between a trotter and a cantering animal. During the summer there will be ample opportunity for those who love horticulture to take exercise in the flower and vegetable garden attached to the institution, and such as wished might be assigned little plots of ground whose management and produce should exclusively belong to them. Looking for a moment from the therapeutics to the economics of the matter, I can see no reason why the house might not rely largely upon itself for at least its summer vegetables and its fruit—if the poorer patients were permitted to pay part of their dues, when they so elected and the exertion was not too much for them, by taking care of the grounds. Another admirable means of exercise will be found in rowing. Our Island must have a good substantial boat-house, containing a good-sized barge for excursions and several pleasure-boats pulling two or three pair of sculls each ; perhaps, eventually, a pair of racing-boats for such of our guests as were well enough to manage a club. Bath-houses for the convenience of those who love a plunge or a swim will be indispensable— affording facilities for a species of summer exercise which nothing can replace.

In winter and summer the bath must be our principal reliance for promoting that vigorous action of the excretory system which with healthy nutrition is our great aim in treating the patient.

Quackery has to so great an extent monopolized the therapeutic use of water, and so much arrant nonsense has been talked in that pure element's name, that we are in danger of overlooking its wonderful value as a curative means. It is one of the most powerful agents at the command of the practitioner, and should no more be trifled with than arsenic or opium. Used by a blundering, shallow-pated empiric it may be worse than useless—may do, as in many cases it has done, incalculable mischief to a patient. In the hands of a clear-sighted, experienced, scientific man, who administers it according to well-known laws of physiology and therapeutics, it is an inestimable remedy, often capable of accomplishing cures without the assistance of any other medicine, and, indeed, where all other has failed. Many of the forms in which it is applied at water-cures well deserve adoption by the more scientific practitioner. Among these *the pack* occupies a front rank. During Mr. Edgerton's month of diminution we use this with him daily. Its sedative effect, when given about three and a half P.M., just after the second dose of bromide of potassium, is exceedingly happy—seeming, as I have heard a patient remark, "to smooth all the fur down the right way"—removing entirely the excessive nervous irritability of the opium-craving, and often affording the patient his only hour of unbroken sleep during the twenty-four. Its tendency to promote perspiration makes it a most effective means for restoring the activity of the opium-eater's skin, and this benefit will be still further increased if it be followed by sponging down the body with strong brine at a temperature as low as the patient can healthily react from, concluding the operation with a vigorous hand-rubbing administered by the attendant until the skin shines. This same salt sponge is a most invigorating bath to be taken immediately on rising. Another excellent bath in use at water-cures, of value both for its tonic and sedative properties, is "*the dripping sheet*," in which a sheet like that used in the pack, of strong muslin and ample size, is immersed in a pail of fresh water at about 70° F., and, without wringing, spread around the standing patient so as to envelop him from neck to feet, the attendant rubbing him energetically with hands outside it for several minutes till he is all aglow. In cases where great oppression is felt at the epigastrium—that *corded* sensation so much complained of by opium-eaters during their earlier period of abandonment, and that peculiar self-consciousness of the stomach which follows in the track of awakening organic vitality—the greatest relief may be expected from "*hot fomentations.*" This is the well-known "hot and wet external application" of the regular practice, and consists of a many-folded square of flannel wrung out of water as hot as the skin can bear, and laid over the pit of the stomach, with renewals as often as the temperature perceptibly falls.

The symptom of cerebral congestion—a chronic sense of fullness in the head—is often very simply alleviated by placing the patient in "*a sitz*" or hip-bath, with the water varying from 70° to 90° F. *Enemata* will constantly be found of service where the torpidity of the bowels is extreme. Not only so, but in cases where the liver is beginning to re-assert itself, and its tremendous overaction sends down such a supply of bile as to provoke inversion of the pylorus, an enema may often act sympathetically beyond that portion of the intestine actually reached by it, and change the direction of the intestinal movement, so as to convert the deadly nausea excited by the presence of bile in the stomach into a harmless diarrhea which at once removes the cause of the suffering. Of the value of foot-baths I need not speak, and to the hot full-bath I must now make reference as the most indispensable agent in ameliorating the sufferings of one who has completely abandoned the drug.

When Mr. Edgerton's dose has reached as low an ebb as $\frac{1}{2}$ grain of morphia he abandons the drug entirely. In my *Harper's Magazine* article I have fully depicted the sufferings which now ensue—as fully, at least, as they can be depicted on paper—though that at the best must be a mere bird's-eye view. During the period of diminution he has endured considerable uneasiness and distress, but these have been trifling to compare with the suffering which he must endure for the first few days and nights, at least, after total abandonment. Universal experience testifies that although the previous period of diminution greatly shortens and softens the sufferings to be endured after giving up opium altogether, the descent from $\frac{1}{2}$ grain of morphia to none at all must involve a few days at least of severe suffering, which nothing borne during the diminution at all foreshadows.

In my *Harper's* article I have said :

" An employment of the hot bath in what would ordinarily be excess is absolutely necessary as a sedative throughout the first week of the struggle. I have had several patients whom during this period I plunged into water at * 110° F. as often as fifteen times in a single day—each bath lasting as long as the patient experienced relief."

Science and experience have thus far revealed no other way of making tolerable the agonizing pain which Mr. Edgerton now endures. This pain is quite inconceivable by the

* On some occasions, by repeated additions from the hot faucet as the temperature of the water in the bath-tub fell, I have raised the bath as high as 120° F. without causing any inconvenience to the patient. Most bath-tubs—all in our own city houses—are too capacious, and too broad for their depth. To prevent cooling by evaporation the tub should be just the width of a broad pair of shoulders and about two feet deep.

ordinary mind.   It can not be described, and the only hint by which an outsider can be let into something like an inkling of it is the supposition (which I have elsewhere used) that pain has become *fluidized,* and is throbbing through the arteries like a column of quicksilver undergoing rhythmical movement.   If the arteries were rigid glass tubes, and the pain quicksilver indeed, there could not be a more striking impression of ebb and flow every second against some stout elastic diaphragm whose percussion seems the pain which is felt.   This is especially the case along the course of the sciatic nerve and all its branches, where the pulse of pain is so agonizing that the sufferer can not keep his legs still for an instant.   There is occasionally severe pain of this kind in the arms also, but this is very rare.   The suffering which usually accompanies that of the legs is a maddening frontal headache, and a dull perpetual ache through the region of the kidneys, described as a sensation of "breaking in two at the waist;" nausea, burning, and constriction about the epigastrium, and intense sensitiveness of the liver—besides general nervous and mental distress which has neither representative nor parallel.

All these symptoms are instantaneously met and for the time being counteracted by the hot-bath.   When the patient gets tired of it, and it temporarily loses its efficiency from this cause, great advantage may be gained by substituting either the Russian bath or the common box vapor-bath, with an aperture in the top to stick the head out of, and a close-fitting collar of soft rubber to prevent the escape of the steam.

I must here refer to another means of alleviation, concerning which I can not bear the witness of personal experience, but which has been highly recommended to me.   Even this brief sketch of treatment would be imperfect without at least a mention of it, and if it possesses all the value claimed for it by persons of judgment who have reported it to me, it will form an indispensable part of our apparatus on Lord's Island. This is an air-tight iron box of strongly-riveted boiler plates, with a bottom and top fifteen feet square and sides ten feet high ; thick plate-glass bull's-eyes in each side sufficiently large to light the interior as clearly as an ordinary room ; and a cast-iron door, six feet in height, shutting with a rubber-lined flange, so that all its joints are as air-tight as the rest of the box. Inside of the box, in the centre, stands a table, suitable for reading, writing, draughts, cards, chess, or games of any similiar kind, with comfortable chairs arranged around it corresponding in number to the people who for an hour or two could comfortably occupy the room. In one side of the box is a circular aperture connecting with an iron tube, which in its turn is joined to a powerful condensing air-pump outside, and on the other side is a pressure gauge with its index inside the box. Sufferers from severe neuralgic pain being admitted, the air-tight door is shut ; they seat themselves, and the condensing pump is set in motion by an engine until the gauge within indicates a pressure of any amount desired. I am told that the severest cases of neuralgia have found instantaneous and thorough relief by the addition of six or eight atmospheres to the usual pressure of air upon the surface of the body. There is no reason why the condensation might not be continued to twenty or more, the increased density causing no uneasiness to those within the box, the same equilibrium between internal and outward pressure that exists everywhere in the air being maintained here. Persons who have made trial of this apparatus speak of the cessation of their pain as something magical; say they can feel it leaving them with every stroke of the pump ; and although as yet we may not be able to offer a scientific explanation of the relief afforded, we can not fail to see its applicability to the case of the reforming opium-eater. If it does all that is claimed for it, it probably acts both mechanically and chemically—the pressure, even though imperceptible from its even distribution, affecting the body like the shampooing, kneading action of an attendant's hand, and the vastly increased volume of oxygen which it affords to the lungs and pores accelerating those processes of vital decomposition by which the causes of many a pain, but especially that of our patient, are to be removed.

The shampooing just referred to, and previously mentioned as forming one process in the Russian bath, is another means of relief constantly in use while the patient is going through his terrible struggle. Our attendants upon Lord's Island are picked men. We do not proceed on the principle in such favor among most of our public institutions, asylums, water-cures, and the like, of procuring the very cheapest servants we can get, and thinking it an economical triumph to chuckle over if* we can manage our patients with the aid of subordinates at twenty dollars a month. We know that in the long run it will pre-eminently *pay* to engage the best people, and we pay the wages which such deserve— wages such as will ensure their quality. Our attendants are selected from the strongest, healthiest, best-tempered, most cheerful-minded, kindest-hearted, most industrious and faithful men and women we can find — people not afraid of work and indefatigable in it—people who understand that no office they can perform for the sick is de-

* This is all that the "canny" business men who compose the managing boards of some of the first asylums in this country permit the heads of the institutions to offer those who must for twenty-three hours of the twenty-four be responsible for the moral and physical well-being of a class of patients (the insane) who require, above all others, wisdom, tact, benevolence, courage, fidelity, and the highest virtues and capacities in those who attend them.

grading or menial, and who will not object, when the patient needs it, to lift him like a baby and rub him vigorously with their hands for an hour at a time. This rubbing our patient often finds the most heavenly relief, not only right after a bath, but at any hour of the day or night. There is, therefore, no hour of either during which Mr. Edgerton can not procure this means of relief from some servant upon duty. Applied to the back and legs especially, it is a sovereign soother for both the opium-eater's acute pain and that *malaise* which is only less terrible. In very severe cases it may be necessary to rub the patient for many consecutive hours, and in such cases it may be necessary either to assign an attendant to the patient's sole care, or, better yet, to have several attendants relieve each other in the manual labor. If the patient could afford and desired it, I should approve of his having his own private servant during the worst of the struggle to perform this labor for him, with the distinct understanding, however, that he was to be private only in the sense of devoting himself to this patient solely, and to receive all his orders from the head of the institution. The expense of such an arrangement would be trifling compared with the amount and intensity of agony which it would save, and in a case of no longer standing than Mr. Edgerton's need last only through the first fortnight or so after abandoning the drug.

Another most important means of alleviation is the galvanic bath. House's patent is an excellent apparatus for the purpose ; convenient in shape and size, comfortable, not easily deranged, affording a variety of simple and combined currents, adjustable so as to pass the current either through the whole body or along almost any nervous tract where it is especially wanted for the relief of local suffering like that of the opium sciatica, and manageable by any intelligent

child who has ever watched attentively while it was getting
put into operation.   Many a sufferer who seems quite a discouraging subject under the dry method of administering
galvanism responds to it at once transmitted through a bath,
and in any case this is a no less beneficial than delightful
way of using it.   The skin is so much better a conductor
when wet, and the distribution by water so uniform, that in
most cases it may be pronounced the best way.

The Turkish bath I have seen used with excellent result
during the earlier days of suffering.   It will seem almost incredible to any one who has taken a Turkish bath for other
purposes, and knows the tax which it seemed to inflict upon
his nervous system for the first few minutes after entering
the heated chamber and till profuse perspiration came to his
relief, when I say that I have seen a man brought to the
bath in that almost dying state of prostration some pages
back described as belonging to the acute attack of opio-
mania, at once subjected to the temperature of 130° F., and
in ten minutes after to thirty degrees higher, not only without
rapidly sinking into fatal collapse, but with a result of almost
immediate and steady improvement.   To my own great surprise his pulse began getting fuller, slower, steadier, and in
every way more normal from the moment that the attendant
laid him down upon his slab.   When he came in he was
obliged to be carried in the arms of his friends like an infant ; his pulse one minute was 140, the next 40–60, or entirely imperceptible, and when fastest alarmingly thready ;
his countenance was corpse-like, he breathed nine or ten
times a minute, and his general prostration so utter that he
could scarcely speak even in a whisper.   He stayed in the
bath an hour and a quarter, in a streaming perspiration for
the last forty minutes, and much of the time sleeping sweetly.
He came out walking easily without assistance, and in the

cool anteroom fell asleep again upon the lounge, not to wake for an hour longer. This one bath entirely broke up the attack. He kept on improving, and with the aid of beef-tea was well enough to go to business in a week. The value of the bath in treating Mr. Edgerton at present will be greatest when he suffers most severely from acute neuralgic pains in the legs and back, especially if the efficiency of the hot full-baths and vapors seem temporarily suspended through frequent use. His own feelings are the best criterion of its worth at any given time. It operates very differently on different people and in different conditions of the system. To some persons it is less debilitating than the use of hot water, and others, myself among the number, find it so excessively disagreeable from the apoplectic sensation it produces in their heads, and the difficulty of breathing which they suffer from it, that nothing but a discovery that it was the *only* means in their particular case of relieving sufferings like those of opium would induce them to enter it. Many persons profess to like it as well as the Russian (which, singularly enough, in no case have I ever known to produce the disagreeable feeling in head or lungs), and it certainly ranks with the foremost alleviatives of the opium suffering—the agonizing rythmical neuralgia of which I have spoken usually becoming magically lulled within two minutes from the time of entering the first heated chamber, and ceasing altogether as soon as the perspiration becomes thoroughly established. At Lord's Island our Turkish bath-room will immediately adjoin our Russian, and the temperature being supported by pipes from the same boiler which furnishes vapor to the other, will be no heavy addition to our expense in the way of apparatus. I don't know whether it is necessary to tell any body that the Turkish bath is merely an exposure of the naked body (with a wet turban around the head) to a dry heat varying from 110° F. to a temperature hot enough to cook an egg hard—followed by ablutions and shampooings somewhat similar to those of the Russian bath.

As it is our aim to *cure* the opium-eater by bringing to bear upon his most complicated of all difficulties every means which has proved effectual in the treatment of any one of its particulars, however caused in other instances, we ask no questions of any appliance regarding its nativity, but take from the empiric whatever he has stumbled on of value as freely as the worthiest discoveries of the philosopher from him. There have been various attempts to erect into a *pathy* every one of the applications we have already mentioned, and I shall close this brief outline of our therapeutic apparatus at Lord's Island with one more valuable method of relief and cure whose enthusiastic discoverers (or rather adapters) have outraged etymology worse than the regular practice by trying to build on their one good thing an entire system under the title of "Motorpathy."* The *"Movement Cure"* contains some very good ideas, which, like many of the Hydropathists', ought to be taken up by Science, in whose hands and their proper place they can do fine service.

As we have found in the case of shampooing, a great deal of the suffering of any part can be taken out by giving it something else to do. A portion of the good done by rubbing an aching leg is no doubt accomplished by setting the nerve at work upon the sensations of pressure and of heat and so diverting it from that of pain, but another portion is probably due to the fact of motions producing changes, in the nature

* I see that some scholar has lately got hold of them and forced them to respect philological canons by kicking the mongrel out of their dictionary and calling themselves *Kinesipathists*, instead of the other Græco-Latin barbarism.

of mechanical and chemical decompositions, in the substance of the tissue ; thus by a well known physiological law summoning a concentration of the nervous forces to the particular part. Nature is thus accelerated in her action there, and as that action is always toward cure (so long as life and hope exist), the nerves of the part are reinforced to act sanely. To be weak is to be miserable—to be strong is to be free from pain—thus the nerve's returning vigor eliminates its suffering. The fresh blood that is pumped into the part by motion brings about another set of ameliorating changes of more especial importance where the pain is caused by a local lesion instead of rather being sympathetic with the whole systematic debility. Whatever be our theory, the tenet that motion relieves pain, as a tenet, is as old as the " *back-straightening*" process used in some shires by the British turnip-hoers who on coming to the end of their rows lie down and let the rest of the women in the field walk over their toil-bent spine and cramped dorsal muscles, while as a fact it is as old as pain itself.

On Lord's Island, therefore, we have a room fitted up with apparatus intended to give passive exercise to every part of the body which the pain of abandoning opium is especially likely to attack.

Mr. Edgerton is suffering extremely, about the close of the third day after his last $\frac{1}{2}$ grain dose of morphia, from the agonizing rythmical neuralgia of which I have spoken, throbbing from the loins to the feet ; and although with good effect we have given him galvanism, shampooing, baths of several kinds, and a number of internal remedies, still, wishing to keep each of these appliances fresh in its potency, we make a change this time to the "movement-room." He is stripped to his shirt, dressing-gown, and drawers, and laid on his back along a comfortable stuffed-leather settee, running quite through whose bottom are a number of holes about four by three and a half inches. These holes are occupied by loose-fitting pistons which play vertically up through the cushion—lying level with it when at rest, and when in motion projecting about two inches above it at the height of their stroke. Motion is secured to them by crank connection with a light shaft running beneath the settee, revolved by a band-wheel, which in its turn connects by a belt with the small engine outside the building, by which all the drudgery of the house is performed. Mr. Edgerton is adjusted over the holes so that, in coming up, the pistons, which are covered with stuffed leather pads, strike him alternately on each side of the spine, from about the region of the kidneys to just beneath the shoulder-blade. The shifting of a lever throws the machine into gear, and for the next five minutes, or as long as he experiences relief, the artificial fists pummel and knead him at any rate of speed desired, according to the adjustment of a brake. This process over, if he still feels pain in the lower extremities, his foot is buckled upon an iron sole which oscillates in any direction according to its method of connection with the power, from side to side, so as to twist the leg about forty-five degrees each way, up and down, to imitate the trotting of the foot, or with a motion which combines several. A variety of other apparatus gives play to other muscles; but I have said enough to show the idea of its *modus operandi.* The passive exercise thus afforded is an admirable substitute for that active kind which in his first few days of deprivation the intensity of his agony often incapacitates him from taking. I have seen men at this period almost bent double from mere pain, and hobbling when they attempted to walk like subjects of inflammatory rheumatism. Their debility also is often so great as to prevent exercise, especially when the characteristic diarrhea has been for some days in operation, though different people differ astonishingly in this respect. I knew one case where an opium-eater of three years' habituation to the drug endured in its abandonment every conceivable distress without suffering from debility at all, as may be inferred from the fact that as his only way of making life tolerable he took a walk of twelve miles every morning while going through his trial. The majority, however, suffer not only pain but prostration of the most distressing character—a combination as terrible as can be conceived, since the former will not let the victim remain in one position for a single minute, and the latter takes away all his own control of his motion, so that he seems a mere helpless, buffeted mass of agony—an involuntary devil-possessed, devil-driven body, consciousness at its keenest, will at its deepest imbecility— almost fainting with fatigue, unable to limp across the room on legs which seem dislocated in every joint and broken in a thousand places, yet unable to stop tossing from side to side, and writhing like a trodden worm all night, all day, perhaps for weeks.

"Oh!" I have heard the patient say, "would to God this made me *tired!* healthily tired, so that I could fall into a minute's doze!"

The apparatus I have just been describing meets this want. Sometimes while the leather and iron fists are pegging away and pummelling him at their hardest, he falls asleep on the machine! It has done for him all that he had not the strength to do for himself—tired him healthily.

The remedies I have mentioned are capable of indefinite combinations. The head of an institution like Lord's Island will want them all, although any one given case may not require all of them. In the hands of a thoroughly scientific, skillful man, they form an armory of means with which such an amount of good can be done as beggars our imagination. Combined with the most faithful attention to the patient's diet—the establishment of healthful nutrition, so that as fast as those abnormal matters which have been clogging the system get cleared away by Nature's relentless processes of decomposition, fresh material may be soundly built up into the system to replace the strength which the fatal stimulant feigned—combined with vigilant, tender, patient nursing— the means described are probably, in many cases, adequate of themselves to restore any opium-eater who is salvable at all.   Still, brief as this sketch is, and so far from making any pretensions to be an exhaustive treatise for the guidance of the profession, I should fail of presenting even a fair outline of the treatment which an unusually wide experience with opium-eaters has convinced me to be the true one, did I not add to the above a few words regarding the medicinal agents which are of value during the month of peculiar trial through which Mr. Edgerton is now passing.

It is scarcely necessary to premise that no such thing as a succedaneum for opium is comprehended in the list of these agents.   Any drug which would so nearly accomplish for the opium-eater what opium accomplishes that he would not miss the latter, must be nowise preferable to opium itself. Such a drug must be able to prevent the decompositions which cause the suffering ; to continue that semi-paralysis of the organic functions in which opium's greatest fascination exists, a paralysis leaving the cerebral man free to exhaust all the vitality of the system in pleasant feelings, lofty imaginings, and aerial dreams, without a protest from the gauglionic man who lies a mere stupefied beggar without any share in the funds of the partnership wherewith to carry on the business of the stomach and bowels and heart, the kidneys and lungs and liver.   It must be a drug that can prevent

the re-awakening of the nutritive and excretory processes—
for it is these whose waking, seeing how late in the day it is,
clamoring at the confusion in which they find affairs and at the
immense quantity of behind-hand work suddenly thrown on
them, together with that re-sharpening of long-dulled sensation by which the clamor comes into consciousness loud as
the world must be to a totally deaf man suddenly presented
with his hearing, which constitute the series of phenomena
which we call pain. No! there is no such thing as a substitute for opium, save—more opium or death. And I do not
know that I need say "*or.*"

Still, there are many alleviatives by which the suffering
may be rendered more endurable—by which now and then
our patient may be helped to catch a few moments of that
heavenly unconsciousness which makes the nervous system
stronger to fight the battle out to its blessed end—by which
processes of Nature may be slowed when they get too fiery-
forceful for human courage to endure, or accelerated when
the pull seems likely to be such a long one as to kill or drive
mad through sheer exhaustion. I have spoken of bromide
of potassium. This in connection with the pack may in
many cases wisely be continued throughout the whole progress of the case, and often hastens the restoration of general
nervous equilibrium by many days, removing to a very perceptible degree that *hyperaesthesia*, that exaggerated sensation of all the natural processes normally unconscious, which
continues to rob the sufferer of sleep long after acute pain is
lulled. The greatest variety of opinions prevails upon the
subject of cannabis and scutellaria. The principal objection to the cannabis lies in two facts. First, it is very
difficult to obtain any two consecutive specimens of the same
strength, even from the same manufacturer. Second, in its
gum state it is exceedingly slow of digestion, and unlike opium

not seeming to affect the system at all by direct absorption
through the walls of the stomach, it is very slow in its action ;
the dose you give at 4 P.M. may not manifest itself till 9 or
even midnight, and even then may still move so sluggishly that
you get from it only a prolonged, dull, unpleasant effect instead of a rapid, favorable, and well-defined one.   If it is
given in the form of a fluid extract or tincture, its operation
can be more definitely measured and counted on, but the
amount of alcohol required to dissolve it is sufficient often
to complicate its effects very prejudicially, while in any case
the immense proportion of inert rubbish, gum, green extractive, woody fibre, and earthy residuum is so great as to
be a severe tax on the digestive apparatus—often seriously
to derange the stomach of the well man who uses it, and
much more the exquisitely sensitive organ of the opium-eater.
I might add a third objection—the fact that its effects vary
so wonderfully in different people—but the physician can
soon get over that by making his patient's constitution in the
course of a few experiments with the drug the subject of his
careful study.   Both its lack of uniformity and its difficulty
of exhibition may be nullified by using the active principle.
It has been one of the *opprobria medicinæ* that in a drug
known to possess such wonderful properties so little advance
has been made toward the isolation of the alkaloid or resin-
oid on which it depends for its potency.   I have for years
been endeavoring to interest some of our great manufacturing
pharmaceutists in the attainment of a form—condensed, uniform, and portable—which should stand to cannabis in the
same relation which morphia bears to opium.   I believe that,
in collaboration with my friend Dr. Frank A. Schlitz (a young
German chemist of remarkable ability and with a brilliant
professional career before him), I have at last attained this
desideratum.   I have no room or right here to dwell upon

this interesting discovery further than to say that we have obtained a substance we suppose to bear the analogy desired and to deserve the title of *Cannabin*. If further examination shall establish our result, we have in the form of grayish-white acicular crystals a substance which stands to cannabis in nearly the same proportional relation of potency as morphia to opium, and this most powerful remedy can be given as easily and certainly as any in the pharmacopœia. If we are successful we shall ere long present it to the medical profession. With all the objections that prejudice cannabis now, I have still witnessed repeated proofs of its great value in lulling pain and procuring sleep, when all other means had failed with the reforming opium-eater, in doses of from one drachm to five of fluid extract or tincture (in some rare cases even larger), administered twice a day. Like opium it is only secondarily a soporific, and to produce this effect it should be given three or four hours before the intended bed-time. Then the earliest effect will be a cerebral stimulus, sufficient to divert the mind from the body's sufferings during day-light, and the reaction will come on in time to produce slumber of a more peaceful and refreshing character—more nearly like normal sleep in a strong, energetic constitution fatigued by healthy exertion, than that invoked by any drug I know of.

It may sometimes be necessary, when the pain has become so maddening and been so protracted, to save the brain from the delirium of exhaustion (or even as I have known to happen, *death*) by procuring sleep for half an hour at any cost save that of a return. The most interesting patient and noble man whose sufferings compose the text and prompted the writing of my *Harper's Magazine* article, died just as it was going to press through the exhaustion of a brain that had no true sleep for months. To avoid such a termination, sleep must be had at any cost, and even the danger attending chloroform or ether must be risked, though I need not point out the necessity of pre-eminent wisdom, and the constant personal presence and watchfulness of symptoms, in the physician during the time that the anæsthetic is inhaled. Of ether as much as three or four ounces may be inhaled during a single evening without much danger, if the precaution of alternating the inspirations from a saturated handkerchief with those of pure atmospheric air be carefully attended to. Chloroform is much more risky, and almost always tends to derange the stomach for several days after its use, still its action is certain in some cases where ether fails even to obscure sensation, and must be resorted to. A single ounce per evening, inhaled with rather longer intervals between whiffs, need not be a perilous dose, and in my experience has often conferred magical relief. Nitrous oxide is too transient to be of much use, but to the extent of twenty or thirty gallons may be used with pleasant effect and about five minutes of alleviation.

Very different from these powerful agents is the humble, much-neglected *scutellaria*. It has been repeatedly pronounced inert, but is beyond all question a minor sedative of charmingly soothing properties, giving sleep, as I have sometimes witnessed, out of the very midst of intolerable rythmical neuralgic suffering—in one case the first sleep the patient had enjoyed since leaving off opium. It may be given with impunity in much larger doses, but on those constitutions with which it has any effect at all a table-spoonful is usually efficacious about ten minutes after its exhibition in the form of fluid extract. Lupulin, valerian, valerianate of zinc, and hyoscyamus (or with a much less tendency to derange the stomach, *hyoscyamin* in $\frac{1}{10}$ grain doses) all have their value in the less violent cases or toward the close of the

struggle. Capsicum, in the five grain doses earlier mentioned, may often be relied on to counteract the tendency to
frightful dreams arising from the exquisitely irritable state of
the stomach in which the opium-habit leaves its victims.

Our object with Mr. Edgerton during the month of struggle has been to assist Nature in eliminating the obsolete matters of the system by all the excretory passages as preparative to the rebuilding of his system on a healthy plan by new
material. During most of the time he has suffered from a
profuse and weakening diarrhea, but this we have not
checked nor retarded, because it was Nature's indispensable
condition precedent to the new man. His perspiration has
been profuse, and that we have assisted for the same reason
by every means in our power—all our baths and rubbings, our
galvanism and medicine so far as used, have favored to the
utmost the activity of his skin. Our repeated hot-baths have
greatly relaxed him ; he may have come to the end of his
month so weak that he could not walk a quarter of a mile if
his life depended on it. No matter. This, however alarming at first sight, is good practice. The more rapidly he has
become relaxed, the further and the further we have banished
pain, from whose presence a state of *tension* is inseparable.
We have not injured him. It is astonishing to any one accustomed to dealing only with the prostration of ordinary
disease to see to what an extremity the opium-eater will bear
to be reduced—what an extent of muscular debility he will
even thrive under. If we look at him closely, we will see
through all his pallor a healthy texture of skin—in all his
languor a *soundness* of vital operation which stands to his account for more valid strength than if he could lift all the
weights of Dr. Winship. Unless the opium-disease is complicated with some serious organic difficulty it is safe to carry
on the process of relaxation as long as it relieves pain until

the patient has just enough strength left to lift his eyelids.
We have kept him up with the constant, faithful administration of beef-tea—half a tea-cupful, by slow sips, every hour
or hour and a half that he was awake during day or night,
but never rousing him for any purpose whatever if he showed
any inclination to sleep.   The nurse who does that when an
opium-eater is going through his struggle should be discharged without warning.   Sleep for ten minutes any time
during this month is worth to nutrition alone more than a
week's feeding.

At the end of the month Mr. Edgerton can sleep with tolerable soundness for half an hour—even an hour at a time,
and the sum of all his dozes amount to about four hours out
of the twenty-four.   He is still nervous, though the painful
tigerish restlessness is gone.   The pangs of his opium-neu-
ralgia are also gone — or re-appear at long intervals, and
much mitigated, to stay but a few minutes.   He is in every
respect on the upward grade.   When his sleep becomes decidedly better, so that most of his night, despite frequent
wakings, is consumed in it, he enters on an entirely different
stage of his treatment.   We stop pulling him down.   We
begin toning him up.

To the description of this process I need devote but little
room.   It consists in a gradual cooling of the temperature of
his baths—a substitution of the more bracing and invigorating for one after another of the relaxing and soothing forms
of treatment.   The hot full-bath is discontinued almost entirely, and we replace it by the use of a couple of pailfuls
of water at 65–75, doused over the patient ; or "the flow," in
which the water spreads through a fan-shaped faucet like a
funnel with its sides smashed flat, and falls over his shoulders ;
or the salt sponge—all followed by vigorous towel and hand-
rubbing until the skin is in a healthy glow.   The pack

we still employ, wringing the sheet out of water as near the natural temperature as he can comfortably and at once react from. It is an admirable means of equalizing the circulation of our patient and soothing his remaining nervous irritability. We encourage his being in the open air and sunshine as much as is compatible with the season and the weather, and favor his taking exercise in every unexhausting way possible. His appetite will by this time take care of his nutrition without much nursing, but we must listen to its caprices and provide it with every thing it thinks it would like. Our sedative medicines may in all likelihood be safely discontinued, and very little indeed of any kind be given him save tonics. In my experience, and that of all others to whom I have recommended them, the very best and most universally to be relied on at this stage are quinine, nux vomica tincture, and pyrophosphate of iron, together with last, but most important of all, our invaluable stand-by, beef-tea. This may be made more palatable to the fastidious palate which has become palled by a steady month or two of it, by a few whole cloves and shreds of onion, but most people relish its delicious meaty flavor quite as well when it is simply made by chopping lean rump into pieces the size of dice, covering them with cold water in the proportion of about three pints to two pounds, letting the whole stand a couple of hours to soak in a saucepan, then drawing it forward upon the range, where it will gently simmer for ten minutes, and salting and pouring it out just as it comes up to a brisk boil. If the meat be just slightly browned on both sides (not broiled through, remember) before being chopped, the flavor of the tea is to many tastes still more exquisite. Beef-tea should be on the range, ready for patients in our house who need it, at all hours of the day and night, and all the year round. The whole cookery of our establishment must be of the very best. There is

no greater mistake than that existing in most sanitary institutions—stinting in the larder and the kitchen.   The best
meats, the most skillful, delicate cookery, the freshest of vegetables and fruit, the ability to tempt the capricious palate by
all sorts of savory little made dishes—these should always
characterize the table of a place where food has to do so
much as with us in replacing the fatal supports of the narcotics and stimuli.   It will be noticed that neither here nor in
my mention of tonics have I referred to alcoholic stimulants.
The omission has been intentional.   My entire experience
has gone to prove that the use of alcohol in any form with
opium-eaters undergoing cure is worse than useless, almost
invariably redoubling their suffering from loss of opium, and
frequently rendering the craving for a return to their curse
an incontrollable agony.   I therefore leave it entirely out,
alike of my pharmacopœia and my bill of fare.

A few final words about the attractions of the Island.
Besides the amusements earlier mentioned, I propose that our
perfected scheme shall contain every thing necessary to make
the social life in-doors a delightful refuge, to all far enough
advanced to take pleasure in society, from the dejection and
introversion peculiarly characteristic of opium's revenges.
This comprehends a suite of parlors where ladies and gentlemen can meet in the evening on just the same refined and
pleasant terms that belong to an elegant home elsewhere ;
furnished with piano to dance to, play, or sing with ; first-
class pictures as fast as our own funds, aided by donations
and bequests, can procure them for us—but bare wall or handsome paper or fresco rather than any daub to fill a panel ;
fine engravings in portfolios ; cosy open fire-places ; unblem-
ished taste in furniture and carpets ; in fine, an air of the
highest ideal of a private family's handsomest assembling-
room.   I propose a billiard-room with a couple of tables—so

neatly kept that both ladies and gentlemen can meet there to enjoy the game, a reading-room with the best papers and magazines and a good library, both to be enjoyed by guests of either sex ; a smoking and card-room for the gentlemen. I propose to have our engine before mentioned do the work of taking our invalids up and down stairs by a lift, like those in use in some of our best hotels, so that the highest rooms may be practically as near the baths, the dining and social apartments, and as eligible as any of the lower ones. And if feasible, I suggest that some at least of the rooms be arranged in small suites or pairs, so as to admit of a well daughter, son, sister, parent, wife, or brother coming to stay with any invalid who needs their loving presence and nursing.

I have thus given as clear an outline as I can of my idea what such an institution as we have so often talked over ought to be, and described a method of treatment which has been successful wherever I have had the opportunity even to approach its realization. For its perfect realization an institution especially devoted to the noble work is a *sine qua non*. If the publication of this letter shall call to our aid in its establishment, by awakening to a sense of its necessity, any of our vigorous, public-spirited countrymen, I am sure we may live to see it flourishing on a sound basis and doing an incalculable amount of good which shall make mankind wonder how so many generations ever lived without it since opium began to scourge the world. I shall then, too, be even more indebted to you than I am now for the courtesy which has afforded so large a space in your book to

Your Friend,

FITZ HUGH LUDLOW.

P

Valuable Standard Works

FOR PUBLIC AND PRIVATE LIBRARIES,

Published by HARPER & BROTHERS, New York.

For a full List of Books suitable for Presentation, see HARPER & BROTHERS' TRADE-LIST *and* CATALOGUE, *which may be had gratuitously on application to the Publishers personally, or by letter enclosing Five Cents.*

HARPER & BROTHERS *will send any of the following works by Mail, postage prepaid, to any part of the United States, on receipt of the Price.*

NAPOLEON'S LIFE OF CÆSAR. The History of Julius Cæsar. By His Imperial Majesty NAPOLEON III. Volumes I. and II. now ready. Library Edition, 8vo, Cloth, $3 50 per volume; Half Calf, $5 75 per volume. *Maps to Vols. I. and II. sold separately. Price* $1 50 each, NET.

MOTLEY'S DUTCH REPUBLIC. The Rise of the Dutch Republic. A History. By JOHN LOTHROP MOTLEY, LL.D., D.C.L. With a Portrait of William of Orange. 3 vols., 8vo, Cloth, $10 50; Sheep, $12 00; Half Calf, $17 25.

MOTLEY'S UNITED NETHERLANDS. History of the United Netherlands: from the Death of William the Silent to the Synod of Dort. With a full View of the English-Dutch Struggle against Spain, and of the Origin and Destruction of the Spanish Armada. By JOHN LOTHROP MOTLEY, LL.D., D.C.L., Author of "The Rise of the Dutch Republic." 4 vols., 8vo, Cloth $14 00; Sheep, $16 00; Half Calf, $23 00.

WOOD'S HOMES WITHOUT HANDS. Homes Without Hands: Being a Description of the Habitations of Animals, classed according to their Principle of Construction. By J. G. WOOD, M.A., F.L.S., Author of "Illustrated Natural History." With about 140 Illustrations, engraved by G. Pearson, from Original Designs made by F. W. Keyl and E. A. Smith, under the Author's Superintendence. 8vo, Cloth, Beveled Edges, $4 50; Full Morocco, $8 00.

ALCOCK'S JAPAN. The Capital of the Tycoon: A Narrative of a Three Years' Residence in Japan. By Sir RUTHERFORD ALCOCK, K.C.B., Her Majesty's Envoy Extraordinary and Minister Plenipotentiary in Japan. With Maps and Engravings. 2 vols., 12mo, Cloth, $3 50.

ALFORD'S GREEK TESTAMENT. The Greek Testament: with a critically-revised Text; a Digest of Various Readings; Marginal References to Verbal and Idiomatic Usage; Prolegomena; and a Critical and Exegetical Commentary. For the Use of Theological Students and Ministers. By HENRY ALFORD, D.D., Dean of Canterbury. Vol. I., containing the Four Gospels. 944 pages, 8vo, Cloth, $6 00; Sheep, $6 50; Half Calf, $8 25.

ALISON'S HISTORY OF EUROPE. FIRST SERIES: From the Commencement of the French Revolution, in 1789, to the Restoration of the Bourbons, in 1815. [In addition to the Notes on Chapter LXXVI., which correct the errors of the original work concerning the United States, a copious Analytical Index has been appended to this American edition.] SECOND SERIES: From the Fall of Napoleon, in 1815, to the Accession of Louis Napoleon, in 1852. 8 vols., 8vo, Cloth, $16 00; Half Calf, $34 00.

BURNS'S LIFE AND WORKS. The Life and Works of Robert Burns. Edited by ROBERT CHAMBERS. 4 vols., 12mo, Cloth, $6 00.

BARTH'S NORTH AND CENTRAL AFRICA. Travels and Discoveries in North and Central Africa: Being a Journal of an Expedition undertaken under the Auspices of H.B.M.'s Government, in the Years 1849-1855. By HENRY BARTH, Ph.D., D.C.L. Illustrated. Complete in 3 vols., 8vo, Cloth, $12 00; Half Calf, $18 75.

BEECHER'S AUTOBIOGRAPHY, &c. Autobiography, Correspondence, &c., of Lyman Beecher, D.D. Edited by his Son, CHARLES BEECHER. With Three Steel Portraits and Engravings on Wood. In 2 vols., 12mo, Cloth, $5 00; Half Calf, $8 50.

BRODHEAD'S HISTORY OF NEW YORK. History of the State of New York. By JOHN ROMEYN BRODHEAD. First Period, 1609-1664. 8vo, Cloth, $3 00.

CARLYLE'S FREDERICK THE GREAT. History of Friedrich II., called Frederick the Great. By THOMAS CARLYLE. Portraits, Maps, Plans, &c. 6 vols., 12mo, Cloth, $12 00; Half Calf, $22 50.

CARLYLE'S FRENCH REVOLUTION. History of the French Revolution. Newly Revised by the Author, with Index, &c. 2 vols., 12mo, Cloth, $3 50.

CARLYLE'S OLIVER CROMWELL. Letters and Speeches of Oliver Cromwell. With Elucidations and Connecting Narrative. 2 vols., 12mo, Cloth, $3 50.

CHALMERS'S POSTHUMOUS WORKS. The Posthumous Works of Dr. Chalmers. Edited by his Son-in-Law, Rev. WILLIAM HANNA, LL.D. Complete in 9 vols., 12mo, Cloth, 13 50; Half Calf, $29 25.

JOHNSON'S COMPLETE WORKS. The Works of Samuel Johnson, LL.D. With an Essay on his Life and Genius, by ARTHUR MURPHY, Esq. Portrait of Johnson. 2 vols., 8vo, Cloth, $4 00.

CLAYTON'S QUEENS OF SONG. Queens of Song: Being Memoirs of some of the most celebrated Female Vocalists who have performed on the Lyric Stage from the Earliest Days of Opera to the Present Time. To which is added a Chronological List of all the Operas that have been performed in Europe. By ELLEN CREATHORNE CLAYTON. With Portraits. 8vo, Cloth, $3 00; Half Calf, $5 25.

COLERIDGE'S COMPLETE WORKS. The Complete Works of Samuel Taylor Coleridge. With an Introductory Essay upon his Philosophical and Theological Opinions. Edited by Professor SHEDD. Complete in 7 vols. With a fine Portrait. Small 8vo, Cloth, $10 50; Half Calf, $22 75.

CURTIS'S HISTORY OF THE CONSTITUTION. History of the Origin, Formation, and Adoption of the Constitution of the United States. By GEORGE TICKNOR CURTIS. Complete in two large and handsome Octavo Volumes. Cloth, $6 00; Sheep, $7 00; Half Calf, $10 50.

DAVIS'S CARTHAGE. Carthage and her Remains: Being an Account of the Excavations and Researches on the Site of the Phœnician Metropolis in Africa and other ad jacent Places. Conducted under the Auspices of Her Majesty's Government. By Dr. DAVIS, F.R.G.S. Profusely Illustrated with Maps, Woodcuts, Chromo-Lithographs, &c. 8vo, Cloth, $4 00.

GIBBON'S ROME. History of the Decline and Fall of the Roman Empire. By EDWARD GIBBON. With Notes by Rev. H. H. MILMAN and M. GUIZOT. A new cheap Edition. To which is added a Complete Index of the whole Work, and a Portrait of the Author. 6 vols., 12mo (uniform with Hume), Cloth, $9 00; Half Calf, $19 50.

www.ingramcontent.com/pod-product-compliance
Lightning Source LLC
Chambersburg PA
CBHW021457210326
41599CB00012B/1040